98-HHE-1

CRYPTOSPORIDIUM IN WASTEWATER: OCCURRENCE, REMOVAL, AND INACTIVATION

by:

Randi M. McCuin
Jennifer L. Clancy, Ph.D.
Clancy Environmental Consultants, Inc.

2005

WERF

The Water Environment Research Foundation, a not-for-profit organization, funds and manages water quality research for its subscribers through a diverse public-private partnership between municipal utilities, corporations, academia, industry, and the federal government. WERF subscribers include municipal and regional water and wastewater utilities, industrial corporations, environmental engineering firms, and others that share a commitment to cost-effective water quality solutions. WERF is dedicated to advancing science and technology addressing water quality issues as they impact water resources, the atmosphere, the lands, and quality of life.

For more information, contact:
Water Environment Research Foundation
635 Slaters Lane, Suite 300
Alexandria, VA 22314-1117
Tel: (703) 684-2470
Fax: (703) 299-0742
www.werf.org
werf@werf.org

This report was co-published by the following organization. For nonsubscriber sales information, contact:

IWA Publishing
Alliance House, 12 Caxton Street
London SW1H 0QS, United Kingdom
Tel: +44 (0) 20 7654 5500
Fax: +44 (0) 20 7654 5555
www.iwapublishing.com
publications@iwap.co.uk

Library of Congress Catalog Card Number: 2004103014

IWAP ISBN: 978-1-84339-718-2

Clancy Environmental Consultants, Inc.

The research on which this report is based was funded, in part, by the United States Environmental Protection Agency through Cooperative Agreement No. CR-825237-01 with the Water Environment Research Foundation (WERF). Unless an EPA logo appears on the cover, this report is a publication of WERF, not EPA. Funds awarded under the Cooperative Agreement cited above were not used for editorial services, reproduction, printing, or distribution.

ACKNOWLEDGMENTS

The authors acknowledge the assistance of Dr. Zia Bukhari, now with American Water Works Service Company in Belleville, Illinois, who was a team member when this project started. Drs. George di Giovanni and Walter Quintero-Betancourt at Texas A&M University provided assistance with the infectivity assays. The project could not have been done without the assistance of 10 utilities that provided samples regularly for analysis to complete the occurrence survey. The City of St. Albans Wastewater Treatment Facility provided weekly samples for three years in method development; this was greatly appreciated. Finally, the authors thank the project subcommittee for their insights and assistance throughout the project, and the WERF project manager, Bonnie Bailey.

Report Preparation

Principal Investigators:
Jennifer L. Clancy, Ph.D.
Clancy Environmental Consultants, Inc.

Randi M. McCuin
Clancy Environmental Consultants, Inc.

Project Subcommittee

Michael J. Arrowood, Ph.D.
Centers for Disease Control and Prevention

Christopher Crockett, Ph.D., P.E.
Philadelphia Water Department

Richard E. Danielson, Ph.D.
BioVir Laboratories

Colin R. Fricker, Ph.D.
CRF Consulting

Sydney Munger
Parametrix, Inc.

Paul A. Rochelle, Ph.D.
Metropolitan Water District of Southern California

Michael Royer
United States Environmental Protection Agency

Stephen A. Schaub, Ph.D.
United States Environmental Protection Agency

Water Environment Research Foundation Staff

Director of Research: Daniel M. Woltering, Ph.D.
Project Manager: Bonnie Bailey

ABSTRACT AND BENEFITS

Abstract:

As our knowledge of established and emerging pathogens in water has expanded, so has the need to examine their occurrence, distribution, risk to humans, and control through treatment beyond chemical disinfection of drinking water. The increased need for wastewater reuse has refocused attention on wastewater treatment processes and their ability to reduce the numbers of pathogenic organisms to acceptable levels. A public health concern with wastewater is the potential for transmission of infectious agents that may be present in human and animal feces.

This study focused on one pathogen, *Cryptosporidium parvum*, and its occurrence in wastewater. To conduct an occurrence study, it was first necessary to develop methods for recovery of *Cryptosporidium* oocysts from wastewater matrices. Because of the differences in matrix composition from raw sewage to tertiary effluents, different methods for recovery and concentration were developed based on matrix quality. A single method was developed for raw sewage and primary influents and a second for secondary and tertiary effluents. The methods for raw influents and primary, secondary, and tertiary effluents were used in a survey of *Cryptosporidium* occurrence at 10 wastewater treatment plants in the United States over 15 months. To determine if oocysts found in wastewater samples represented a public health risk, cell culture methods examined infectivity of recovered oocysts.

Despite considerable effort to develop a method to isolate *Cryptosporidium* oocysts in biosolids, trials performed with oocyst recoveries greater than 40% were successful only with biosolids containing low percent solids (<10%) and with small amounts of sample (≤5 g). More research is needed to develop a method to recover oocysts in biosolids with higher solid content.

Benefits:

- Develops and characterizes methods for recovery and enumeration of *Cryptosporidium* oocysts from wastewater samples.
- Shows that significant difficulties remain in using cell culture and molecular methods for determining oocyst infectivity from wastewater matrices.
- Describes how oocyst adhesion to particulates affects recoveries in the wastewater matrices and shows that more research is needed to develop and optimize *Cryptosporidium* recovery methods with biosolids.
- Demonstrates that *Cryptosporidium* oocyst removal cannot be done reliably even in drinking water studies unless high seeding levels are used in carefully designed and conducted experiments.

Keywords: Cryptosporidiosis, cell culture, wastewater, biosolids, pathogens, public health, survey

TABLE OF CONTENTS

LIST OF TABLES

LIST OF FIGURES

LIST OF ACRONYMS

ACS	American Chemical Society
AHBSS	Acidified Hank's Balanced Salts Solution
ALCOSAN	Allegheny County Sanitary Authority
ANOVA	One-way Analysis of Variance
ASTM	American Society for Testing and Materials
Caco-2	Human Colonic Adenocarcinoma
CC-QSD	Cell Culture-quantitation Sequence Detection
CFC	Continuous Flow Centrifuge
Ct	Cycle Threshold
DABCO	Diazabicyclo-[2.2.2]-octane
DAPI	4'-6'-diamidino-2-phenylindole
DI water	Deionized Water
DIC	Differential Interference Contrast
EDTA	Ethylenediamine Tetracetic Acid
FA	Fluorescent Antibodies
FBS	Fetal Bovine Serum
FDM	Focus Detection Method
FITC	Fluorescein Isothiocynate
g	Relative Centrifugal Force
HBSS	Hank's Balanced Salts Solution
HCT-8	Human Ileocaecal Adenocarcinoma
HSP	Heat Shock Protein
HV	High Volume
ICR	Information Collection Rule
IMS	Immunomagnetic Separation
IO	Indigenous Oocysts
IONICs	Intact Oocysts No Infection Controls
mAb	Monoclonal Antibody
mRNA	Messenger RNA
NaHMP	Sodium Hexametaphosphate
ntu	Nephelometric Turbidity unit
PBS	Phosphate-buffered Saline
PBST	Phosphate-buffered Saline with 0.01% Tween 20
PCR	Polymerase Chain Reaction
PCTE	Polycarbonate Track-etched
ppm	Parts Per Million
QA/QC	Quality Assurance/Quality Control
RFLP	Restriction Fragment Length Polymorphism
rRNA	Ribosomal Ribonucleic Acid
RT-PCR	Reverse Transcription-polymerase Chain Reaction
SCA	Standing Committee of Analysts
SD	Standard Deviation
sp gr	Specific Gravity
TAMU	Texas A&M University

U.S. EPA	United States Environmental Protection Agency
UV	Ultraviolet
WWTP	Wastewater Treatment Plant

EXECUTIVE SUMMARY

Chemical disinfection of drinking water was once considered sufficient for reducing the risk of infection from pathogenic organisms. However, as our knowledge of established and emerging pathogens in water has expanded, so has the need to examine their occurrence, distribution, risk to humans, and control through treatment. The increased need for wastewater reuse has resulted in attention now being focused on wastewater treatment processes and their ability to reduce the numbers of pathogenic organisms to acceptable levels. A public health concern with wastewater is the potential for transmission of infectious agents that may be present in human and animal feces. Depending on the diseases in the contributing communities, sewage can contain varying numbers of pathogenic organisms including viruses, bacteria, helminthes, and protozoa.

Cryptosporidium is a significant concern to water suppliers worldwide. This protozoan parasite forms highly resistant oocysts that can survive in moist environments for extended periods. The oocysts are highly resistant to environmental pressures and are difficult to remove in water treatment because of their small size (4–6 μm). *Cryptosporidium* oocysts are resistant to conventional chlorine based disinfectants, and oocysts that escape the coagulation/filtration process can remain viable in the distribution system. Dozens of outbreaks of cryptosporidiosis related to sewage contamination of drinking water have occurred worldwide, with deaths resulting.

Animals and humans are reservoirs for this parasite, which enters the environment through shedding of fecal material. Dozens of species harbor *Cryptosporidium* oocysts, including mammals (cattle, horse, rodent, deer, dog, cat), marsupials (kangaroo), birds, reptiles, and fish species. As such, there are many routes for this parasite to enter the environment including natural runoff (nonpoint sources), runoff from agriculture, effluents from industries such as meat processors, wastewater effluents, and combined and sanitary sewer overflows.

Following treatment, sewage effluents are discharged in receiving waters, and these waters in turn may be used as sources of water for human consumption after treatment, or for agricultural or recreational purposes. The presence of small numbers of viable oocysts in potable waters can pose a significant threat of infection to individuals or communities that consume such water. At the start of this project there were no standard methods in existence for determining the presence of *Cryptosporidium* in wastewater influent or effluent samples. Thus, it was important to develop and optimize methods that would allow improved evaluation of wastewater treatment performance with respect to *Cryptosporidium* inactivation or removal.

In 1997, the U.S. Environmental Protection Agency (U.S. EPA) began development and testing of a new *Cryptosporidium* method for water, which was a significant improvement over previous methods. This method is U.S. EPA Method 1622: *Cryptosporidium* in Water by

Filtration/IMS. This current project used Method 1622, but modified it specifically to handle difficult wastewater matrices. Three separate methods were developed for wastewater matrices. Because of the differences in matrix composition from raw sewage to tertiary effluents, different methods for recovery and enumeration of oocysts were developed based on matrix quality. A single method was developed for raw sewage and primary influents that uses centrifugation to concentrate 500 mL samples. A second method for secondary and tertiary effluents uses the Envirochek high volume (HV) capsule for sample concentration. Both methods rely on immunomagnetic separation (IMS) to isolate oocysts from debris, followed by antibody staining and microscopy for identification and enumeration. The inclusion of *Color*Seed, an internal positive control in each sample, provided information on method performance. These methods were used in a survey of *Cryptosporidium* occurrence in 10 wastewater treatment plants (WWTPs) over 15 months. Summary data from the 289 samples in the survey are presented below.

Table ES-1. Summary of *Color*Seed Recoveries, Turbidities, and Number of Samples with Indigenous *Cryptosporidium* Oocysts in Wastewater Samples Collected from 10 WWTPs across the United States from May 2002 to July 2003.

Matrix (*n*)	Volume Analyzed (L)	Range of Turbidities (ntu)	Range of *Color*Seed Recoveries (%)	*Color*Seed Recoveries (%) (Mean ± SD)	Concentration of IO[2] Detected (# / L)	Samples positive for IO (%)
Raw Influent (95)	0.5	34.3–700	0–83.8	26.1 ± 17.7	<2–24	29.5
Primary Effluent (84)	0.5	4.08–327	0–78.8	33.0 ± 17.9	<2–86	45.8
Secondary Effluent (94)	4.2–22.7	0.75–19.3	0–62.6	25.0 ± 16.6	<0.1–40.8	58.5
Tertiary Effluent (16)	22.7–131.3	0.75–2.62	13.1–65.7	48.8 ± 14.5	<0.008–0.226	18.8

1. ntu, nephelometric turbidity unit.
2. IO, indigenous oocysts.

Methods to isolate *Cryptosporidium* oocysts from biosolids or fecal material used by other researchers were examined in this study, however oocyst recovery rates were generally poor. These techniques included slow-speed centrifugation, and density gradient flotation. Modifications to the U.S. EPA method to recovery Ascaris ova from biosolids also resulted in low oocyst recoveries. Although considerable effort was given to developing a method to isolate *Cryptosporidium* oocysts from biosolids, trials with oocyst recoveries greater than 40% were accomplished only with biosolids with low percent solids (<10%) and with a small amount of sample (≤5 gm). When the level of solids increased in this matrix, oocyst recoveries plummeted. To overcome this, the amount of sample analyzed was decreased but resulting recoveries were still poor. The authors feel that will the current technologies available recovering oocysts from biosolids is tedious and difficult with low recoveries often resulting. Because of the obstacles encountered in our method development trials we did not included biosolids in the occurrence survey.

Another objective of the occurrence study was to examine the removal of oocysts across the treatment process. It stands to reason that the concentration of oocysts should be continually decreasing across the treatment process, from raw influent to final effluent (secondary or tertiary) on a per volume basis. However, this was not apparent from the data presented. Figure ES-1 shows the oocyst concentration per liter at Plant #3 in Alabama. In some months the concentration of oocysts recovered is higher in secondary effluent than in raw influent or primary effluent (May, October, December, March). Because *Cryptosporidium* cannot reproduce outside of an animal host, the concentration of oocysts cannot increase through the treatment process. There may be several reasons for this observation. First, the low overall levels of oocysts detected in each wastewater matrix make it difficult to show significant log_{10} differences. Another factor is the potential to recover oocysts more efficiently from cleaner water matrices; hence greater numbers of oocysts were recovered from effluent samples than from raw or primary influents samples. *Cryptosporidium* oocysts present in raw influents or primary effluents may be embedded in biological particles as they enter the plant and become dispersed during the treatment process and are more readily recovered in the cleaner wastewater matrices. Similar trends were noted for other plants, with one exception. The Vermont tertiary treatment plant was an anomaly, as no oocysts were recovered from 32 of the 33 samples. Eleven oocysts were observed in a raw influent sample in February 2002.

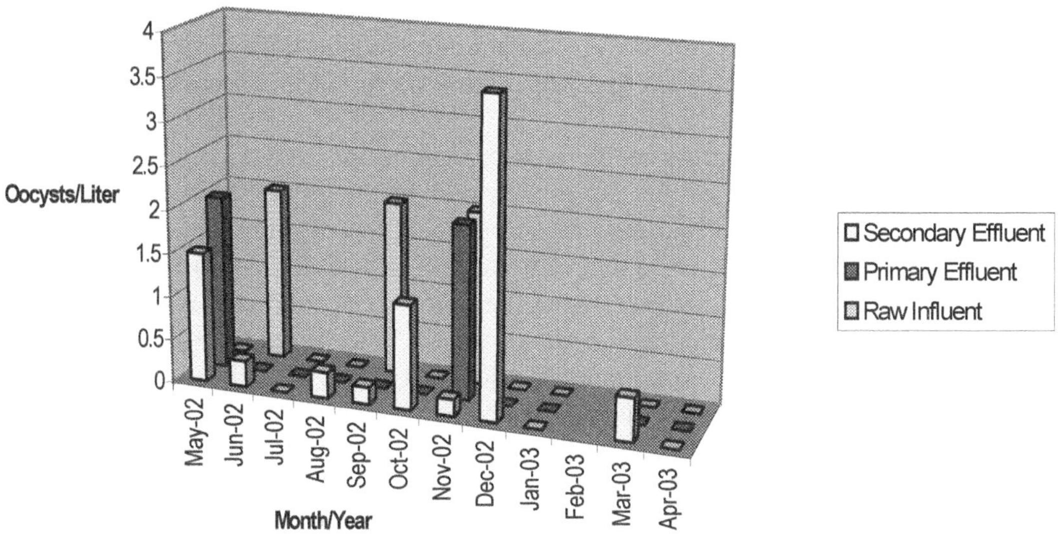

Figure ES-1. Indigenous Oocyst Concentrations at Alabama Plant #3 from May 2002 through April 2003.

To determine if oocysts found in wastewater samples present a public health risk, it is necessary to know if they are infectious. The assay currently used to detect oocysts in water is a microscopic method, and both live and dead oocysts are detected, and appear similar morphologically in most instances. There is no way to know if the oocysts observed present a potential public health threat; the current method is limited in this aspect. To make a risk determination, additional testing to determine infectivity of recovered oocysts is needed.

Cell culture and molecular methods were employed to examine the infectivity of oocysts recovered from various wastewater matrices. Using cell culture-sequence quantitation detection (CC-QSD; LeChevallier et al., 2003) results were difficult to interpret. When oocysts were seeded into autoclaved secondary effluent samples at levels from 10 to 100, the method could not detect infectious oocysts. It is possible that the oocysts used in the seeding study had a low infectivity rate. However, control samples using the same oocyst preparation (pure oocyst preparations, no wastewater sample) responded appropriately. Another reason for results showing no infectious oocysts may be because of interferences present in the matrix. In raw influent, primary, secondary and tertiary effluents tested using the CC-QSD method, results were obtained showing very high levels of infectious oocysts present. Again, the wastewater sample concentrates were heat inactivated to "kill" any potentially present indigenous oocysts so that infections detected would only be from seeded oocysts. These erroneous results may indicate a positive interference present in all samples tested or more probable an amplification of nontarget DNA.

Results using the focus detection method (FDM) showed different infectivity results from results obtained using the CC-QSD method. In control oocyst suspensions analyzed in duplicate at concentrations of 50 and 200 oocysts, low infectivity rates were obtained when only one infection focus was detected in one of the monolayers seeded with approximately 200 oocysts. Because of this low infection rate, no infection sites were detected in secondary wastewater samples collected from a Pennsylvania plant and seeded with oocysts at two different concentrations. However, in seeded secondary effluent samples collected at the California wastewater facility indigenous infectious oocysts were detected at a level of 2–6 oocysts/10 L. No additional infection foci were detected in the samples seeded with purified oocyst suspension collected at this site because of the low level of infectivity noted in the control samples.

Cell culture and molecular methods to detect infectious oocysts in wastewater are still works in progress. Although molecular techniques such as the CC-QSD assay have the potential to provide enumeration results of oocysts present, elimination of interferences and nontarget organisms must be optimized to ensure that only target DNA is detected in the amplification portion of the assay. The results obtained using the FDM were encouraging despite low infectivity rates obtained with the control oocysts. Again, interferences must be eliminated to ensure that oocyst can invade the monolayers without apoptosis resulting. However, detection of infectious oocysts was only one objective of the project. The major objectives that were planned were achieved. The research developed the first methods designed to specifically recover oocysts from wastewater. In addition, the 15-month survey at 10 U.S. WWTPs provides the most comprehensive dataset available on *Cryptosporidium* occurrence in wastewater.

CHAPTER 1.0

EXPERIMENTAL PLAN AND APPROACH

1.1 Overview

Treatment of drinking water was once considered sufficient for reducing the risk of infection from pathogenic organisms. However, as our knowledge of established and emerging pathogens in water has expanded, so has the need to examine their occurrence, distribution, risk to humans, and control through treatment. In addition, the increased need for wastewater reuse has resulted in attention now being focused on wastewater treatment processes and their ability to reduce the numbers of pathogenic organisms to acceptable levels. A public health concern with wastewater is the potential for transmission of infectious agents that may be present in human and animal feces. Depending upon the diseases in the contributing communities, sewage can contain varying numbers of pathogenic organisms including viruses, bacteria, helminths and protozoa. The species of protozoan parasites that are considered to be of major health significance to humans include *Entamoeba histolytica, Giardia intestinalis* (= *lamblia* = *duodenalis),* and *C. parvum* (Straub et al., 1993).

In humans, cryptosporidiosis is caused by *C. parvum* and *C. hominis*, with the most common symptoms being profuse, watery diarrhea, abdominal cramping, nausea, vomiting, and low-grade fever. In most well nourished, immunocompetent individuals, infection may last between 2 and 12 days and is usually self-limiting, however, in some patients, the infection may continue for two weeks or longer and occasionally fluid replacement therapy may be required. In poorly nourished individuals, especially children, diarrhea may last for longer than 3 weeks. Because excessive fluid losses occur in these cases, oral or parenteral rehydration often becomes necessary. As no effective chemotherapeutic treatment is currently available for cryptosporidiosis, the duration of diarrheal illness depends on the patient's immune status. In patients with congenital or acquired immune deficiencies or malnourished individuals, infection can be considerably prolonged, resulting in malabsorption, severe dehydration, and even death. In individuals with acquired immunodeficiency syndrome, *Cryptosporidium* infections are first established in the ileum or jejunum and can spread throughout the gastrointestinal tract.

In the United Kingdom, more than 8,000 cases of cryptosporidiosis have been reported annually and data from the Public Health Laboratory Services indicate that *C. parvum* infections were the fourth most common cause of gastrointestinal infections diagnosed in England and Wales (Anon., 1990). In the United States, data presented in *Morbidity and Mortality Weekly Reports (MMWR,* 1996*)* by the Centers for Disease Control and Prevention indicate that although cryptosporidiosis is not notifiable in all states, 2470 cases were reported in 2002 and 3,111 cases were reported in 2003. In addition to domestic sources, effluents from agricultural and commercial enterprises (e.g., cattle markets, abattoirs, etc.) can also be responsible for releasing large numbers of *Cryptosporidium* sp. oocysts into sewage.

The first recorded outbreak of waterborne cryptosporidiosis was associated with contamination of a well water supply with oocysts derived from sewage (D'Antonio et al., 1985). Examination of stool specimens from 79 residents who consumed water from the well source indicated that gastrointestinal illness was 12 times greater than in the surrounding communities that consumed water from a different water source. Since then, numerous waterborne outbreaks of human cryptosporidiosis have been reported. Because of their potential for causing disease in large numbers of individuals—for example, the Milwaukee outbreak, which affected an estimated 403,000 individuals (MacKenzie et al., 1994)—the waterborne route of oocyst transmission has become a focus of considerable research effort over the last decade. Excretion of large quantities of oocysts by infected animals or humans and direct fecal contamination of water can increase the risk of waterborne human cryptosporidiosis. In addition, disposal of oocyst-contaminated animal excrement on to grazing land and the subsequent contamination of pasture can lead to increased risks of infection for grazing animals. The high prevalence of *Cryptosporidium* spp. in neonatal animals and the excretion of large numbers of oocysts ensures a high level of environmental contamination, which, when followed by periods of heavy rainfall, can lead to surface run-off and contamination of water sources with oocysts. A summary of some waterborne outbreaks of human cryptosporidiosis, with sewage as the implicated source of contamination, is presented in Table 1-1.

Sewage effluents have the potential to cause contamination of the aquatic environment with oocysts, and as the presence of small numbers of viable oocysts in potable waters can pose a significant threat of infection to individuals or communities that consume such water. In a human feeding study of healthy volunteers, the ID_{50} (infectious dose 50%) of the Iowa strain of *C. parvum* oocysts was determined to be 132 oocysts however, infections occurred in this study with doses as small as 30 oocysts (DuPont et al., 1995). It is important to be able to evaluate the presence, removal, and inactivation of *Cryptosporidium* sp. oocysts during sewage treatment processes. Before this study, no standard methods existed for determining the presence, removal, and inactivation for *Cryptosporidium* sp. oocysts that are derived from sewage influent and effluent samples. Following treatment, sewage effluents are discharged into receiving waters and these waters may in turn be used either as sources of water for human consumption after treatment, or for agricultural or recreational purposes. Thus it is important to develop or optimize methods that would allow improved evaluation of sewage treatment processes performance with respect to removing or inactivating the protozoan parasites belonging to the genus *Cryptosporidium*.

1.2 Development of Improved Filtration Technologies for Environmental Waters

Inability to amplify *C. parvum* oocysts reliably by using *in vitro* cultivation procedures has meant that, unlike other organisms of significance to the water and wastewater industries, the presence of individual oocysts has to be determined for samples under evaluation. Source and finished water samples most often contain low numbers of oocysts; therefore large volumes of water may need to be examined. Although wastewater samples have the potential to contain highly variable numbers of these organisms, the level of sediment and particles present in wastewater are significant factors influencing sample collection and processing. As a result, both the detection and sample collection/concentration procedures for wastewater need to be modified from those used in for source and finished waters.

The procedures for isolation and detection of *Cryptosporidium* oocysts consist of three separate stages: (1) sample collection and concentration, (2) separation of oocysts from contaminating debris, and (3) detection of the oocysts (or their contents) (Clancy et al., 1999). In recent years additional steps to determine infectivity of oocysts recovered from environmental matrices using cell culture and molecular techniques have been reported. These techniques show promise for use in the water industry, but no standard protocols have been established among researchers for use among the larger scientific community. There is a need for standardization of cell culture and molecular methods so that accurate determinations of infectious oocysts can be made to assess risk to public health.

Table 1-1. Wastewater-Associated Outbreaks of Human Cryptosporidiosis.

Location	Date	Estimated Number of People Affected (Confirmed Cases)*	Raw Water Source	Suspected Sources of Contamination	Reference (Year)
San Antonio, Texas	May–July 1984	2,000 (47)	Well	Raw sewage	D'Antonio et al. (1985)
Carrollton, Ga.	January–February 1987	13,000	River	Raw sewage and runoff from cattle grazing areas	Hayes et al. (1989)
Berks County, Pa.	August 1991	551	Well	Septic tank effluent, nearby creek	Solo-Gabriele & Neumeister (1996)
Jackson County, Ore. (Talent and Medford)	January–June 1992	15,000	Spring/river	Surface water, treated wastewater, or runoff from agricultural areas	Leland et al. (1993)
Milwaukee, Wis.	January–April 1993	403,000	Lake	Cattle wastes, slaughterhouse wastes, and sewage carried by tributary rivers	Kramer et al. (1996)
Cook County, Minn. (Grand Marais)	August 1993	27 (5)	Lake	Backflow of sewage or septic tank effluent into distribution area, raw water inlet lines, or both	*MMWR* (1996)
Las Vegas, Nev.	January–April 1994	(78)	Lake	Treated wastewater, sewage from boats	Roefer et al. (1996)
Walla Walla, Wash.	August–October 1994	86 (15)	Well	Treated wastewater	Dworkin et al. (1996)
Yorkshire, U.K.	1988	67	Swimming pool	Sewage-contaminated swimming pool	Joce et al. (1991)

A procedure originally intended for collection of *Giardia* cysts from water samples utilizing filtration of large sample volumes (100–1000 L) through yarn-wound filters with a nominal 1–μm pore has been used for collection of wastewater, especially effluent samples, for *Cryptosporidium* sp. analysis. Until June 1997 and the development of U.S. EPA method 1622 (U.S. EPA, 1997), this method—known as the Information Collection Rule (ICR) method (U.S. EPA, 1996)—was the only standard method for analyzing water for the presence of *Cryptosporidium* in the United States. Although no standard wastewater methods have been identified, permutations of the ICR method have been used frequently. The problems in the ICR method—low and highly variable recoveries, high false positive and negative rates, and poor precision and accuracy—have been well documented. The method is technically difficult and recoveries are poor, even in expert laboratories (Nieminski et al., 1995; Clancy et al., 1994).

In the United Kingdom, two methods were proposed for collection/concentration of sewage influents/effluents in the Standing Committee of Analysts (SCA) blue book method (Anon, 1990). For sewage effluents, one SCA method requires filtration of 10–50 L of sewage effluents through a polypropylene cartridge filter at a flow rate of 1.5 L/min. The cartridge filters can be processed according to the SCA method (Anon., 1990), which is very similar to the cartridge filter processing/concentration protocols described in the ICR method. Oocysts are isolated from contaminating debris using a cold sucrose solution of a specified density and concentrated to a minimal volume using centrifugation. The second SCA method advocates screening of 5 L grab sample through a coarse (50–150 μm) filter; samples are concentrated by repeated centrifugation to 10 mL, subjected to sucrose density flotation and concentrated to a minimal volume (0.5–1.0 ml; depending upon the levels of contaminating debris). Both these procedures were derived or suggested approaches rather than protocols developed after experimental validation. However, later reported data have indicated that the use of 100-L filtered samples yield consistently lower numbers of oocysts than 2-L grab samples concentrated by centrifugation alone (Robertson et al., 2000; Bukhari et al., 1997). These investigations also indicated that for some filtered samples it was necessary to perform two or three sucrose density flotation steps before samples were in an acceptable state for microscopy.

In the last decade, various sample collection devices have become available and some have been incorporated into the U.S. EPA Method 1622 for detection of *Cryptosporidium* oocysts. The most user-friendly of these devices require minimal handling, are suited for small sample volumes and have been reported to retain large proportions of the spike dose and yield recoveries in excess of 50%. These filters include the Pall standard Envirochek™ and Envirochek high volume (HV) capsules. The Envirochek filter has a 1-μm polysulfone absolute filter housed in a polycarbonate case, whereas the HV filter material is comprised of polycarbonate track-etch (PCTE) membranes. The effective filtration area of the HV is 1300 cm^2. Water is passed through these filters at 2 L/min, and the retained material is eluted using a weak Laureth-12 solution. Another filtration system validated for use in Methods 1622/1623 is the IDEXX Filta-Max™ (McCuin & Clancy 2003; U.S. EPA, 1998). This filtration/concentration system was developed for the continuous *Cryptosporidium* monitoring of finished water in the United Kingdom. This system is comprised of a filtration device consisting of 60 foam pads compressed to create a nominal porosity of 1 μm. Samples are pushed through the filter at a flow rate up to 4 L/min. Retained particles are eluted from the filter and the filter wash is concentrated using a particle concentrator. This filtration system is equipped with a 3.0-μm-pore polysulfone membrane filter in the base. A magnetic stir bar is placed in the particle concentrator allowing the sample to mix while a vacuum is applied to concentrate the sample. The theory is that *Cryptosporidium* oocysts remain suspended while the sample is concentrated. The sample concentrate is decanted to a conical centrifuge tube; the membrane filter is removed from the particle concentrator base, washed in a minimal volume of elution buffer, which is added to the sample concentrate. *C. parvum* oocysts retained in the sample concentrate are recovered using specific isolation techniques. This system works well with clean matrices; however, its application for turbid waters (clay particles, silt) and wastewater is limited because the filter located in the base of the particle concentrator readily plugs with increased particle levels necessitating the use of multiple membranes. In addition, the manipulations required in changing out membranes can lead to losses of target organisms. A PCTE membrane system (Corning Costar) has given recovery rates as high as 90%; however, the presently used flat bed membrane format does not provide adequate surface area for filtration of high volumes of turbid water (Shepherd and Wyn-Jones, 1996). A technology that has received considerable attention in the research community is the continuous flow centrifuge (CFC). The portable CFC was originally developed for separation of blood cells and had been used for recovery of *Cryptosporidium* oocysts spiked into tap and river water (Zuckerman et al., 1999). This centrifuge can be used not only for *Cryptosporidium* and *Giardia* concentration in water but is used for *Encephalitozoon* species and viruses. The drawback of this technology is that the CFC is not commercially available.

1.3 Separation of Oocysts from Background Debris

Where samples are collected by filtration, an elution step is normally employed, which nonselectively elutes target organisms from the filter along with contaminating debris. Organisms in filter eluants or grab samples are concentrated by centrifugation, resulting in concentration of *Cryptosporidium* oocysts as well as large amounts of extraneous material. Flotation procedures such as sucrose (specific gravity [sp gr] 1.18) or Percoll-sucrose (sp gr 1.1) were originally used to separate oocysts from extraneous materials. However, these techniques resulted in loss of oocysts and also concentrated other materials (debris) nonspecifically. Bukhari and Smith (1995) reported that flotation procedures selectively concentrated viable oocysts and noted that oocyst recoveries were dependent upon population viability. Increased exposure to water concentrates has also been reported to reduce oocyst recoveries, which was probably because of increasing proportions of nonviable oocysts (Fricker, 1995). Using conventional isolation procedures, Bukhari and Smith (1995) demonstrated that a water-ether sedimentation procedure was highly efficient for oocyst recoveries and had minimal effects on oocyst viability, as determined by the fluorogenic vital dyes assay. Because diethyl ether extracts lipids from samples, it is only capable of yielding partial purification of samples; however, by using a combination of sucrose flotation followed by water-ether sedimentation, the oocysts lost in the sucrose pellet may be recovered in the pellet remaining after water-ether sedimentation. A disadvantage of this approach is that each sample (especially influents) would require two separate enumeration and viability analyses, although it does provide a better indication of enumeration/viability data than with flotation alone.

Several commercially available immunomagnetic separation (IMS) kits have emerged to specifically isolate oocysts from contaminating debris. These procedures are intended to replace the conventional flotation or sedimentation procedures. IMS uses attachment of specific antibodies to magnetizable particles, and efficient mixing of the particles with the samples to form a bead-oocyst complex. This complex can then be extracted from the remaining debris by application of a magnetic field. Although the concept and procedures are simple, there is evidence to indicate differential performance of commercially available kits in water sample concentrates (Bukhari et al., 1998). In this study, two of the three commercially available IMS kits for *Cryptosporidium* were evaluated, Dynabeads® anti-*Cryptosporidium* (Dynal Biotech ASA, Oslo, Norway) and Crypto-Scan™ IMS (Immucell Corp., Portland, Me) kits. The Dynal IMS kit yielded consistently higher recoveries (Bukhari, 1997). Further investigations with this kit also revealed that the age of oocysts did not appear to influence oocyst recovery (Bukhari et al., 1998; McCuin et al., 2001a), although oocyst suspensions in highly turbid water samples can lead to the recoveries being reduced significantly with all three commercially available kits.

1.4 Methods for Detection/Enumeration of *Cryptosporidium* Oocysts

Routine detection of *Cryptosporidium* oocysts relies on the use of epifluorescence microscopy. Usually samples are either collected on 1–2 μm pore, 25-mm diameter membranes or air dried onto welled glass slides and labeled with a genus specific anti-*Cryptosporidium* monoclonal antibody (mAb) that has been conjugated to a fluorochrome such as fluorescein isothiocynate (FITC). Unfortunately these antibodies are only genus-specific and are unable to distinguish between *Cryptosporidium* species or strains of *C. parvum*.

Another issue is the presence of algal species in environmental samples that cross react with the FITC-mAb. These cross-reacting algal cells can exhibit the same size, shape and fluorescent characteristics of *Cryptosporidium* oocysts thereby making a determination of the object's identity presumptive at best. Isolation of oocysts from sample concentrates using IMS have reduced the numbers of cross-reacting cells present in the detection phase of the assay and the use of the fluorogenic stain, 4'-6'-diamidino-2-phenylindole (DAPI), in Methods 1622/1623 has assisted in confirming *Cryptosporidium* oocysts by the staining characteristics of the nuclear DNA of the sporozoites in presumptive oocysts. The limitation of this detection assay is the inability of the stains to determine viable and infectious oocysts. The inclusion of a viability/infectivity assay seems prudent to definitively measure public health risk associated with the presence of these organisms.

1.5 Molecular Assays for *Cryptosporidium*

Over the last decade several different genus- or species-specific primers have been described for amplification of target sites by using the polymerase chain reaction (PCR; Laxer et al., 1991; Rochelle et al., 1997a). Techniques such as reverse transcription (RT)-PCR or *in vitro* cultivation, in conjunction with PCR, have both improved sensitivity for detection of oocysts and also to accrue information on oocyst viability/infectivity (Rochelle et al., 1997b). These techniques represent a tremendous advancement in the endeavor for sensitive detection of low numbers of oocysts. In most cases specificity of these assays in distinguishing *Cryptosporidium* sp. from other organisms has been demonstrated; however problems of species-specificity have been noted. An evaluation of eight previously reported sequences, targeting six distinct regions of the *Cryptosporidium* genome, failed to differentiate between *C. parvum* and *C. meleagridis* (Champliaud et al., 1998). DNA encoding for the small subunit of ribosomal RNA (18S rRNA) has been sequenced for at least three species of *Cryptosporidium* and approximately 97-98% sequence homology has been noted. Several PCR-based assays have been developed with the 18S rRNA (Johnson et al., 1993; Johnson et al., 1995) but the limit of detection with environmental samples can be compromised. Rochelle et al. (1997a) compared primers with optimized PCR conditions and found that to detect 50 oocysts in environmental samples, two successive rounds of amplification, combined with a hybridization assay, were necessary. The gene encoding a heat shock protein (HSP 70) has been investigated for assessing viability of *Cryptosporidium* oocysts (Stinear et al., 1996; Rochelle et al., 1997b). One of the major limitations encountered with these molecular amplification procedures is their sensitivity to the presence of contaminating debris, especially compounds such as humic/fulvic acids, which are inhibitors of the amplification process.

Recent studies incorporatied various cell culture techniques for detecting *Cryptosporidium* in environmental matrices. Rochelle et al. (1999) demonstrated that oocysts recovered from source water pellets using IMS retained their infectivity. Rochelle et al. (2001) investigated a colorimetric *in situ* hybridization assay for quantification of *C. parvum* infections in cell culture. This assay provides a direct enumerative technique with the specificity of the molecular-based methods. Other researchers have reported the FDM as a reliable method for detection of infectious oocysts in environmental matrices (Slifko et al., 1999; Gennaccaro et al., 2003; Quintero-Betancourt, 2003). Slifko et al. (1999) further demonstrated good correlation of FDM to animal infectivity after oocysts were inactivated using a pulsed bright light device. Rochelle et al. (2002) reported good correlation between cell culture combined with RT-PCR and animal infectivity in oocysts exposed to UV and ozone disinfection. These researchers concluded that cell culture offers the advantage of being able to support the infection of both genotypes whereas animal models can only support the type-2 genotype *C. parvum*. The presence of low numbers of oocysts in environmental samples precludes the use of animal infectivity for quantitative determination of recovered oocyst infectivity. However, low numbers of oocysts in environmental matrices can pose a problem for cell culture methods also. Rochelle et al. (2000a) has reported infection rates of less than 7% with fresh oocysts in human ileocaecal adenocarcinoma (HCT-8) cells. The authors continued to report reduced infectivity rates as the oocysts aged. The low infectivity rates reported by Rochelle et al. (2001) using purified oocyst suspensions demonstrate low sensitivity using cell culture techniques. It is possible that appropriate triggers are not present to induce the oocysts to lyse and invade cell monolayers.

1.6 Summary

There is limited information on the effectiveness with which WWTPs remove or inactivate *Cryptosporidium* spp. oocysts, and data from the few studies which have been published indicate considerable variations in oocyst numbers both in raw sewage and treated effluents. In the existing studies, sewage samples have been collected by either large volume filtration or grab samples and because of inconsistencies in sample collection/concentration procedures and variations in recovery efficiencies of employed methods, it is difficult to compare results. Despite the absence of standardized methods, these studies show that raw sewage can contain large numbers of oocysts and that high variations exist in the oocyst numbers that occur either in a single sewage treatment plant or from different plants. When compared with raw sewage, considerably lower numbers of oocysts were detected in treated effluents; however, this does demonstrate that sewage treatment regimes do not remove oocysts completely.

The oocysts removed from raw sewage are likely to be concentrated in sewage sludge and therefore reduction of environmental contamination with viable oocysts necessitates the optimization of sludge treatment processes for reducing oocyst viability. There is little information on the effect of sludge treatment processes on oocyst viability. Data from one published report suggests that oocyst inactivation primarily depends on the temperatures attained during sludge treatment (Stadterman et al., 1995). These authors reported that the temperatures of thermophilic sewage sludge treatment and sludge pasteurization inactivated oocysts within a few hours, whereas sludge treatment at mesophilic temperatures required between one and several days to inactivate oocysts. To corroborate these initial data and to increase understanding of the efficiency of sewage treatment processes, further work should be conducted with standardized sample collection, oocyst concentration, and oocyst detection methods.

1.7 Research Hypotheses and Objectives

The overall goal of this research was to develop methods for recovery and detection of *Cryptosporidium* oocysts in wastewater matrices. The specific objectives were to

- evaluate and modify existing and emerging methods to survey *Cryptosporidium* removal and disinfection in wastewater treatment systems;
- demonstrate the ability of the proposed method(s) to detect and quantify *Cryptosporidium* oocysts in unit processes of a variety of representative WWTPs; and
- determine the levels of inactivation of *Cryptosporidium* oocysts in unit processes of a variety of representative WWTPs.

An addendum was included during Year 4 to address infectivity. Techniques now exist that allow the inclusion of an infectivity step in environmental waters. We were interested in determining if an infectivity assay could be included in a wastewater method using current techniques reported by other researchers. Inclusion of an infectivity step in the *Cryptosporidium* methods would allow meaningful assessment of infectious oocyst occurrence in wastewater and the potential risks to public health.

1.8 Research Approach

The research initially focused on evaluating existing and emerging methods for detection of oocysts in wastewater. Evaluation of these methods was performed using wastewater from a local WWTP with tertiary treatment of the plant effluent. Because of the variable nature of each wastewater matrix it was necessary to evaluate two sample concentration techniques; centrifugation of the raw sewage and primary effluents and filtration/elution/centrifugation of secondary and tertiary effluents. High particle levels in raw sewage and primary effluents necessitated the use of centrifugation of small volumes (250–1000 mL) for sample concentration. Direct examination of the resulting concentrate proved inefficient so isolation of the oocysts was achieved using IMS. It was discovered during development trials that incorporation of kaolin (hydrated aluminum silicate) during the IMS step improved oocyst recoveries in the presence of fecal fats, oil and grease, and high levels of particles normally found in these samples.

Because of reduced levels of particles in secondary and tertiary effluents it was possible to analyze larger volumes of sample as compared to primary effluents. Different sample collection devices were evaluated for their ability to concentrate 10 L of secondary effluent and to recover the retained oocysts from the filter material. Elution, concentration, isolation, and detection of captured oocysts from depth filters such as the spiral-wound polypropylene cartridge, Filta-Max foam module, and the standard Envirochek capsule did not compare favorably with the recovery of the target organisms from PCTE membrane filters such as the CrypTest™ cartridge, 142-mm flat membrane, and the Envirochek HV capsule filters. Another promising technology for sample concentration—continuous flow centrifugation—was investigated; however, the equipment was not readily available to pursue further recovery trials. The final method for the recovery and detection of *C. parvum* oocysts in secondary and tertiary effluents is a modification of U.S. EPA Method 1622 incorporating a pre-elution step with sodium hexametaphosphate (NaHMP).

During method development trials high variability in oocyst recoveries of triplicate subsamples were noted with each matrix, especially raw sewage and primary effluents. Variability in oocyst recoveries was observed from day to day as well as from subsample to subsample. This variability may be caused by physical characteristics of the matrix, such as particle levels, or to chemical interferences. This was demonstrated on a day when no seeded oocysts were recovered from secondary effluent when dairy waste was dumped into the wastewater stream. *Color*Seed became commercially available during this project. Incorporating this internal control with each sample was a way to assess method performance regardless of variability caused by particulate levels or chemical interferences. This product contains flow cytometry sorted, gamma-irradiated oocysts stained with Texas Red. These organisms are distinguished from indigenous oocysts by the red detected using a standard green filter during epifluorescence microscopy.

Once collection, concentration, isolation, and detection techniques were optimized for oocyst recovery in wastewater, WWTPs from different U.S. locations sent grab or composite samples of their raw sewage, primary ,and secondary effluents (and in some cases tertiary effluents). Samples were collected over 15 months; however no utility participated for the entire 15 months. The results of this occurrence survey are included in this report.

Cell culture methods and molecular methods were employed to examine the infectivity of oocysts recovered from various wastewater matrices. The results of this aspect of the study were limited, as difficulties were encountered with the complex wastewater matrices. Where predetermined numbers of purified oocysts (10, 25, 50, and 100 oocysts) were added to autoclaved secondary wastewater subsample concentrates, no infectious oocysts were detected using the CC-QSD assay. Controls using the same oocyst suspension indicate levels of infectious oocysts routinely observed in this assay. Therefore, these results would appear to indicate chemical interference in the wastewater matrix that either "inactivated" the oocysts or prevented the oocysts from causing infection. In other wastewater matrices, infectious oocyst recovery rates of 0–200,000% were noted when seeded with 289 oocysts using the CC-QSD assay. These erroneous results are most likely caused by the amplification of nontarget DNA. Infectious oocyst results using the FDM were more encouraging although the purified oocysts used in the seeding controls had a very low infectivity rate. However, infectious indigenous oocysts were detected in secondary effluents samples collected from a WWTP in California. These results demonstrate that detecting infectious oocysts in wastewater matrices is difficult and requires more research before this is a standard technique included in wastewater analysis.

CHAPTER 2.0

THE RECOVERY OF *CRYPTOSPORIDIUM* OOCYSTS IN RAW SEWAGE AND PRIMARY EFFLUENTS

2.1 Introduction

Raw sewage and primary effluents contain high levels of particles that make filtration of large sample volumes impossible. Because of the high concentration of particles and the complex nature of the wastewater matrices, some researchers have concentrated their efforts on the use of alternate density gradients for clarification and isolation of oocysts from other debris. These procedures offer nonselective isolation of biological particles and may yield preparations containing high levels of contaminating debris that may occlude target organisms in the detection phase of the assay. Using low spike doses (~120 oocysts/L), Robertson et al. (2000) achieved a mean oocyst recovery of 81% in 50–100 mL of sewage influents. Subsequent isolation/flotation techniques using ether, cold sucrose (1.18 sp gr), or combined ether/cold sucrose yielded reductions in the mean oocyst recoveries of approximately 32%. Using 50 mL–2 L grab samples, other researchers have used the centrifugation/clarification approach for recovery of oocysts in sewage influents; however, recovery rates for methods used were not assessed (Bukhari et al., 1997; Payment et al., 2001; Quintnez-Diaz et al., 2001; Gibson 1998; Dumoutier & Mandra, 1996). We investigated some of these previously employed techniques for isolating *C. parvum* oocysts in raw sewage. Filtration of the raw sewage through coarse muslin to reduce the levels of particles present before centrifugation and direct examination was investigated with minimal success. Nonspecific isolation of biological particles present in 50 mL of raw sewage using cold sucrose (sp gr 1.18) and sodium chloride (sp gr 1.2) resulted in oocyst recoveries less than 20% when the harvested interface was examined.

With the development of IMS kits for the specific isolation of *Cryptosporidium* oocysts from water in the mid-1990s, the research team proposed evaluating this technique to improve oocyst isolation from wastewater. Several companies released IMS kits during the course of our research, and the performance of each in wastewater was evaluated. These procedures are intended to replace the conventional flotation or sedimentation procedures. The IMS technique relies on attachment of specific antibodies to magnetizable particles, efficient mixing of the particles with the samples and subsequent formation of a bead-oocyst complex. This complex can then be extracted from the remaining debris by application of a magnetic field.

Although the concept and procedures are simple, there is evidence to indicate differential performance of these kits in water sample concentrates (Bukhari et al., 1998). The authors reported on two of the three commercially available IMS kits for *Cryptosporidium* evaluated, Dynabeads® anti-*Cryptosporidium* (Dynal Biotech ASA, Oslo, Norway) and Crypto-Scan™ IMS

(Immucell Corp., Portland, Maine) kits. The Dynal IMS kit yielded consistently higher recoveries (Bukhari, 1997). Further investigations with this kit also revealed that the age of oocysts did not appear to influence oocyst recovery, although oocyst suspensions in highly turbid water samples can lead to the recoveries being reduced significantly with all three commercially available kits (Bukhari et al., 1998, McCuin et al., 2001a).

One commercial kit offered a different approach to capturing the oocyst-bead complex by circulating the sample past the magnet several times. The Puri-Max system (Genera Technologies, Ltd.) was developed in conjunction with the Filta-Max system for isolation of *Cryptosporidium* oocysts from finished water. The IMS procedure involved circulating the sample concentrate in the presence of bovine serum albumin and IMS beads through an inline filter and back into the sample reservoir. The inline filter removes larger particles that may plug the steel mesh of the paramagnetic phase. After an appropriate incubation time (allowing the beads to interact with the oocysts) the paramagnetic phase was placed in line and samples continued to circulate. The paramagnetic phase theoretically captures the bead-oocyst complex as it passes by the magnetized steel mesh. The steel mesh is eluted and the captured bead-oocyst complexes recovered. Our modified approach to recover captured oocysts was to subject the Puri-Max IMS concentrate to the Dynal IMS procedure to reduce the IMS concentrate to 100 µL allowing the concentrate to be placed on a single well. This approach was abandoned because particles readily plugged the inline filter as well as the paramagnetic phase. Trials conducted without the inline filter were compromised because particles were so large they clogged the tubing.

Other approaches to capture and concentrate oocysts from wastewater were investigated. IDEXX Corporation (Westbrook, Maine) markets a system for the collection and concentration of oocysts in finished water called Filta-Max. This product is used in the United Kingdom for the Continuous *Cryptosporidium* Monitoring Rule (U.K. Drinking Water Inspectorate, 2000). One of the components of the Filta-Max system is a particle concentrator. A membrane filter (3 µm-pore, 72-mm diameter polysulphone) is placed in the base of a sample concentrator tube (length 30 cm x 6.5 cm inner diameter) and the concentrator tube is screwed into the base. Filter eluant is added to the concentrator tube and the concentrator is placed on a magnetic stirrer. The lid of the concentrator, which contains a rod with a magnetic stirrer attachment, is immersed in the sample and screwed in place on top of the concentrator tube. A piece of Tygon tubing is attached to a valve located on the base and connected to a waste container and vacuum source. The magnetic stirrer is switched on at maximum speed, a vacuum is applied, and the valve on the concentrator base is opened simultaneously. This procedure enables filtration of excess supernatant through the 3-µm membrane filter while retaining large particles and organisms of interest in the concentrator tube. Samples are concentrated by filtration to 20–25 mL and the concentrate is transferred to a 50 mL conical centrifuge tube. The membrane is removed from the concentrator base, transferred to a small plastic bag. Phosphate-buffered saline (PBS) with 0.01% Tween 20 (PBST; 15 mL) is added, and the membrane is hand kneaded to remove organisms from the membrane surface. This wash is also added to the respective centrifuge tube. The resulting sample concentrate is subjected to IMS. Recovered oocysts were detected and enumerated using epifluorescence microscopy. Oocyst recoveries using this technique showed very promising results by the analyst who developed the technique. However, when other analysts in the laboratory tried to perform this technique, oocysts recoveries dropped significantly (data not shown). Further attempts to reproduce the original results were unsuccessful, and this approach was abandoned.

In addition to oocyst recoveries in wastewater matrices, we evaluated oocyst recoveries in the American Society for Testing and Materials (ASTM) substitute wastewater, developed for testing chemical parameters, and deionized (DI) water. DI water was selected to assess oocyst recoveries in the absence of interferences. During initial trials evaluating oocyst recoveries in small volumes (50 mL) of raw sewage and primary effluents using centrifugation followed by IMS it was noted that recoveries of target organisms were highest in the artificial matrix. These findings prompted us to determine if the results were caused by one of the components of the substitute matrix. Kaolin (hydrated aluminum silicate) was selected as the first component to test and when added to concentrated wastewater samples during the IMS step we noted oocyst recoveries rose dramatically. Trials were performed to determine the optimal amount of kaolin to add without inhibiting the performance of the IMS kit. With the addition of kaolin it was necessary to perform additional rinses to the bead-oocyst complex to remove particles that may have occluded oocysts during the microscopy phase. However, these additional rinses did not appear to have an adverse effect on oocyst recoveries.

2.2 Materials and Methods

2.2.1 *Cryptosporidium* Oocyst Spike Dose Source and Enumeration Procedures

2.2.1.1 Viable *C. parvum* Oocysts

C. parvum oocysts were obtained from the Sterling Parasitology Research Laboratory, University of Arizona (Tucson). This isolate was originally isolated from a calf and is referred to as the Iowa strain (Moon & Bemrick, 1981). It has been maintained by passage in neonatal calves. The feces of experimentally infected calves was collected and clarified by using cesium chloride density gradient centrifugation (Arrowood & Donaldson, 1996). Purified oocysts were stored in DI water and antibiotics (gentamicin, 0.1 mg/mL; streptomycin, 0.1 mg/mL; and penicillin, 100 units/mL) at 4°C. The age of oocysts used in these trials was less than 3 months. Stock suspensions were diluted to a concentration of $1x\ 10^3$/mL with DI and stored at 4°C.

Suspensions of *C. parvum* oocysts were enumerated as described in Method 1622 (U.S. EPA, 2001). Briefly, 10 100-µL replicate aliquots were spotted onto individual wells of three-well treated microscope slides (Meridian Diagnostics, Inc., Cincinnati, Ohio, product no. R2206), dried at 42°C for 1-2 h, fixed with absolute methanol, and air-dried for 3–5 min. Anti-*Cryptosporidium* sp. FITC-mAb (Waterborne, Inc., New Orleans) were placed into each well. The slides were placed in a humid chamber and incubated at 35°C for 45 min. Excess FITC-mAb was aspirated, and each well was rinsed with 150 mM PBS, pH 7.2, and aspirated. A weak DAPI solution (0.004 µg/mL in PBS) was placed into each well, incubated at room temperature for 2 min. Excess DAPI solution was removed by aspiration. The wells were rinsed with PBS and dried in the dark at room temperature. Mounting medium (2% diazabicyclo [2.2.2] octane—DABCO—in 60% glycerol-40% PBS) was placed in the center of each well, a coverslip was applied, sealed with clear nail polish and slides were examined using epifluorescence microscopy. Spike doses were used within 24 h of enumeration.

2.2.1.2 *Color*Seed™ Flow Cytometer-Sorted Spike Dose Suspensions

*Color*Seed is a commercially available product from Biotechnology Frontiers Pty, Ltd. (BTF, PO Box 599, North Ryde BC, NSW 1670, Australia). This product contains flow cytometer-sorted, gamma-irradiated *Cryptosporidium* oocysts and *Giardia* cysts that have been permanently stained with Texas Red, a fluorescent red dye. Each organism had a mean concentration of 98–100; the standard deviation (SD) of counts of each organism for approved batches must be less than 2.5. Batches are prepared and have a shelf life of 4 months if stored at 2–8°C.

*Color*Seed was transferred to the test sample (or filter) using the method prescribed by the manufacturer. Briefly, 2 mL 0.05% Tween 80 was added to the contents of a *Color*Seed vial. The vial was capped and shaken vigorously 25 times. The cap was removed and the contents were decanted to the test sample. DI water (3 mL) was added to the vial, the cap replaced, and the tube was again vigorously shaken 25 times. The rinse was added to the test sample and the rinse step was repeated.

Trials were conducted to assess if the protocol prescribed by the manufacturer was adequate in transferring all the *Cryptosporidium* oocysts to the test sample. Tween 80 was added to the vial as prescribed by the manufacturer. Instead of transferring the vial contents to a test sample, the contents were applied to a 13-mm diameter, 1.0-µm-pore PCTE membrane filter housed in a filter housing with a fritted support. The housing was rinsed with 0.05% Tween 80 before applying the sample. A vacuum was applied the filter to draw the fluid through the membrane. The vial was rinsed with DI water as prescribed in the instructions and the rinse was also filtered through the PCTE membrane filter. The retained particles were stained with FITC-mAb using the Petri dish method. Briefly, the Petri dish technique for staining captured organisms involves placing the PCTE membrane filter on a 25µL aliquot of FITC-mAb placed in the center of a small Petri dish cover (60 x 15 mm). The Petri dish was placed in a humid chamber (larger Petri dish with a moist paper towel on the bottom) at 37°C for 1 h. The filter was then placed onto a fritted support and a vacuum was applied to draw the excess FITC-mAb through the membrane. PBS was applied dropwise onto the membrane with the vacuum applied to rinse away excess FITC-mAb. The membrane was placed on a glass slide with 10 µL of mounting medium specified in U. S. EPA Method 1622. Another 10 µL aliquot of mounting medium was placed in the center of the membrane and a glass coverslip was placed on top of the membrane. The mounting medium was allowed to spread the edges of the coverslip and was then sealed with clear nailpolish. Enumeration of recovered oocysts was performed with epifluorescence microscopy.

2.2.2 Preparation of the Standard Specification for Substitute Wastewater

Substitute wastewater was prepared according to the specifications outlined in the ASTM D5905-96. Briefly, beer (<100 calories/12 oz.) was transferred to a container allowing 20% headspace. The container was shaken and allowed to sit refrigerated for 24 h to dissipate the carbonation. After 24 h, the container was removed from the refrigerator and allowed to reach room temperature. A household blender was used to mix 500 mL DI water, 0.4 g of all-purpose, bleached flour, 2.0 g sea salts (Sigma S-9883), 0.08 g kaolin (hydrated aluminum silicate) (Sigma K-7375), 20 mL 0.12% Triton X-100 solution, and 120 mL of the light beer, at the lowest speed for 30 s. The foam was allowed to subside, the mixture transferred to a 2 L container and the volume adjusted to 2 L with DI water. This preparation of the substitute wastewater was used within 24 h.

2.2.3 Dynal IMS Procedure

Each sample concentrate was transferred to a Leighton tube containing 1 mL each of SL-A 10X and SL-B 10X buffers. To each tube, 100 µL of the Dynal Dynabeads® Crypto-beads were added. Samples were incubated for 1 h at room temperature while rotating on the sample mixer at 18 rpm. At the end of the incubation period, the Leighton tube was placed in the MPC-1 with the flat side of the tube towards the magnet. The tube was gently rocked through a 90° angle, tilting from tube top to bottom with the magnet side facing up, for 2 min. The supernatant was decanted; the tube removed from the MPC-1 and 1 mL of 1X SL-A buffer was added to the tube. The tube was gently rocked to resuspend the bead complex and using a Pasteur pipette, the suspension was transferred to a 1.5 mL Eppendorf tube. Another 0.5 mL of 1X SL-A buffer was added to the Leighton tube to remove excess residuals and was transferred to the respective Eppendorf tube. Eppendorf tubes were then placed in the MPC-M, the magnet inserted in place and the MPC-M was rocked through 90° for 1 min. The bead-oocyst complex was collected at the back of the eppendorf tube and the supernatant aspirated. The magnet was removed from the MPC-M and 100 µL of 0.1N HCl was added to each sample and vortexed for a minimum of 30 s. The samples were allowed to stand for 10 min in the upright position, vortexed a further 10 s and the magnet of the MPC-M was inserted into place. The tubes were allowed to stand undisturbed for 2 min. The resulting supernatant was placed in the center of a well of a three-welled microscope slide containing 10 µL of 1 N NaOH. Slides were allowed to dry at 42°C for 30 min, fixed with 25 µL absolute methanol, air-dried, and stained with FITC-mAb. Slides were examined using epifluorescence microscopy.

When kaolin was incorporated in the IMS procedure for raw sewage and primary effluents, the following modifications were made to the above protocol. After the Leighton tube was rocked for 2 min in the MPC-1 it was placed on a flat surface undisturbed for 3 min. The sample was then rocked for another 2 min and the supernatant was decanted. Without removing the Leighton tube from the MPC-1, 10 mL of PBS was added down the side of the tube opposite the captured beads. The Leighton tube was removed from the MPC-1 and gently rocked five times to resuspend beads and entrapped particles. The tube was replaced in the MPC-1 with the flat side toward the magnet. The tube was gently rocked through a 90° angle, tilting from tube top to bottom with the magnet side facing up, for 2 min. The supernatant was decanted; the tube removed from the MPC-1 and 1 mL of 1X SL-A buffer was added to the tube. The IMS procedure was completed as described above. Recovered oocysts were enumerated using epifluorescence microscopy.

2.2.4 Aureon IMS Procedure

Samples subjected to the Aureon IMS protocol (Aureon Biosystems GmbH, Vienna, Austria) were concentrated to approximately 5 mL. The pellet was vortexed and transferred to a Leighton tube for isolation of oocysts from the debris using IMS. A 5 mL rinse of IMS buffer 'A' was used to scour residual sample from the centrifuge tube and was transferred to the Leighton tube. For raw sewage, kaolin (0.75 g) and *Cryptosporidium* A-Beads (100 µL) were added and the tube was placed a sample rotator at 25 rpm for 45 min. For secondary and tertiary effluent samples, *Cryptosporidium* A-Beads (100 µL) from the kit were added to the Leighton tube, tubes were capped, and rotated at 25 rpm for 45 min. Samples were then removed from the mixer and inserted into the magnetic separation device (MagnetOn 4T). The magnet was gently rocked for 1 min returned to the upright position and allowed to stand undisturbed for 3 min. The magnet was then gently rocked for another 30 s and the supernatant was discarded. To resuspend the beads, 1 mL of A-Wash buffer was added and the tubes were gently mixed. Using a Pasteur pipette, the resuspended beads were transferred to a 2.0 mL cryovial with a rounded bottom. The vial was capped and placed in the MagnetOn 4T and rocked gently for 30 s. With the vial remaining in the magnetic device, the wash buffer was gently aspirated using a Pasteur pipette. Tubes were rinsed with a fresh 1 mL aliquot of A-Wash buffer, which was then transferred to the cryovial for another rinse. The bead separation procedure was repeated. Fifty microliters of 0.05 N HCl was added to the cryovial and vortexed for 15 s. The vial was allowed to stand for 30 s and then vortexed for an additional 15 s. The cryovial was placed in the MagnetOn 4T and allowed to stand for 10 s. The IMS concentrate was transferred to a well slide. The acid disassociation step was repeated with the second acidic suspension being placed in the same well as the first. The well was placed on the slide warmer to dry. Recovered oocysts were fixed, stained, and enumerated as previously described.

2.2.5 Epifluorescence Microscopy

A Zeiss Axioskop fluorescence microscope, equipped with a blue filter block (excitation wavelength, 490 nm; emission wavelength, 510 nm) was used to detect FITC-mAb-labeled oocysts at a magnification of 360X. DAPI staining characteristics were observed at 640X total magnification using a UV filter block (excitation wavelength, 400 nm; emission wavelength, 420 nm). A green filter block (excitation wavelength, 546 nm; emission wavelength, LP590) was used for visualization of the Texas Red stain of the *Color*Seed at 640X total magnification. Internal morphology of oocysts was observed by using Nomarski differential interference contrast (DIC) optics microscopy at 640–1600X total magnification.

2.2.6 Optimizing *Cryptosporidium* Oocyst Recoveries in Raw Influents and Primary Effluents

2.2.6.1 Processing of 50 mL Samples Spiked with Oocysts

Eight 50 mL aliquots of each matrix (DI water, substitute wastewater, raw sewage, and primary effluent) tested were prepared in 50 mL conical centrifuge tubes. Six of these tubes were spiked with 100 or 500 oocysts. Each tube was vortexed and concentrated by centrifugation (1050 x *g*; 10 min). Three spiked samples and one unspiked sample were processed and examined directly while the remaining samples were processed for detection using IMS. For direct sample examination, the supernatant was aspirated to approximately 300 μL. The pellet was vortexed and split between three individual wells of a three-welled slide. Samples subjected to IMS for isolation of oocysts were concentrated as described above and the supernatant was aspirated to 10 mL. The concentrate was vortexed to resuspend pelleted material.

2.2.6.2 Oocyst Recoveries in Raw Sewage Using Centrifugation in 250 ml and 500 ml Centrifuge Tubes

Triplicate 500 mL of raw sewage were seeded with approximately 100 *C. parvum* oocysts. Tween 80 (25 mL of 20% solution) was added for a final concentration of 1% to each sample and shaken to thoroughly mix. The content of each tube was concentrated by centrifugation (1500 x *g*; 15 min) and the supernatant was aspirated to 10 mL. The concentrate was transferred to a Leighton tube containing IMS buffers. Oocysts were recovered from the interfering debris in the presence of 0.75 g kaolin. A fourth sample was processed unspiked to determine background levels of target organisms. In addition, four 500 mL of DI water containing Tween 80 (final concentration, 1%) were seeded with approximately 100 oocysts and concentrated using centrifugation and processed as described above. Recovered oocysts were enumerated using epifluorescence microscopy.

Triplicate subsamples (250 mL, 500 mL, and/or 1 L samples) were seeded with approximately 100 oocysts. Tween 80 was added to each sample to yield a final concentration of 1% and each sample was mixed thoroughly. The samples were then concentrated by centrifugation (1500 x *g*; 15 min) in 250 mL conical centrifuge tubes and the supernatant was aspirated to approximately 5 mL. In addition, four 250 mL samples of DI water containing Tween 80 (final concentration 1%) were seeded with approximately 100 oocysts and concentrated using centrifugation. The IMS procedure was carried out as previously described in the presence of 0.75 g kaolin. Unseeded subsamples were processed to determine background levels of the target organisms. IMS sample concentrates were stained and recovered oocysts were enumerated using epifluorescence microscopy.

2.2.6.3 Evaluation of Matrix Interference on IMS Performance

To determine interference of the matrix on the performance of the IMS triplicate volumes (250 mL, 500 mL, and 1 L) of raw sewage containing Tween 80 (1% final concentration) were concentrated using centrifugation (1500 x *g*; 15 min) using 250 mL conical centrifuge tubes. The supernatant was aspirated to 5–8 mL. For samples in excess of 250 mL, additional unconcentrated sample was added to the same tube containing the sample concentrate. The sample was concentrated by centrifugation and the supernatant was aspirated to 5–8 mL. This concentration procedure was continued until the entire sample volume was concentrated. Pellet material was transferred to Leighton tubes and approximately 100 oocysts were seeded to each tube. The IMS procedure was performed in the presence of 0.75 g kaolin, as previously described. Recovered oocysts were enumerated using epifluorescence microscopy.

2.2.6.4 Evaluation of Oocyst Recoveries Using Optimized Method

Raw influent and primary effluent grabs (10 L) were collected on the day of analysis. In initial trials, triplicate subsamples (250 mL, 500 mL, and/or 1 L) were seeded with approximately 100 live oocysts. In addition, a subsample of wastewater at each volume tested was analyzed unseeded to determine background concentration of oocysts in the sample. In later trials, *Color*Seed was also added to each test volume using the transfer protocol described above. An appropriate volume of 20% Tween 80 solution was added to each subsample to yield a final concentration of 1% and was then thoroughly mixed. Each sample was concentrated by centrifugation (1500 x *g*; 15 min) in 250 mL conical centrifuge tubes (Corning Inc., Corning, N.Y.; part no. 430776), the pellet volume recorded and the supernatant aspirated to 5-8 mL. *C. parvum* oocysts in the concentrates were isolated using either the Dynal or Aureon IMS kits as described below; recovered organisms were enumerated using epifluorescence microscopy.

2.3 Results and Discussion

Table 2-1 presents oocyst recovery data in 50 mL of raw sewage, primary effluent, DI water, and substitute wastewater. With a spike dose of approximately 100, mean oocyst recoveries were 38.9 ± 5.2% in DI water when direct examination of the sample concentrate was performed. The mean recovery rose to 47.6 ± 1.0% when the sample concentrate was subjected to IMS using the Dynal system. Direct examination of the substitute wastewater matrix resulted in oocyst recoveries of 60.7± 9.3% and 68.6 ± 6.6% in two trials using a spike dose of approximately 100 oocysts. When a spike dose of approximately 500 oocysts was used the mean oocyst recovery was 59.2 ± 9.0%. In two of three trials conducted with the substitute wastewater the oocysts recoveries dropped to approximately 50% when IMS was employed to isolate oocysts from the interfering debris. Direct analysis of seeded 50 mL raw sewage concentrates yielded mean oocyst recoveries of 21.1± 3.5% and 49.2 ± 5.9% using doses of approximately 100 and 500, respectively. The mean oocyst recovery rose to 39.7 ± 4.3% in raw sewage concentrates subjected to IMS at spike dose concentrations of 100 oocysts. However, at a spike dose concentration of 500 oocysts the mean recovery dropped to 37.3 ± 9.7%. In trials using 50 mL primary effluent, mean recoveries were approximately 20% when samples were examined directly. These mean recoveries rose to 36.6 ± 16.6% and 43.8 ± 24.2% with spike doses of 100 and 500 oocysts, respectively, when sample concentrates were subjected to IMS.

Table 2-1. A Comparison of Detection Methods for *Cryptosporidium* Oocyst Recoveries in Four Matrices Using 50 mL Grab Samples.

Sample Detection Method (no. of replicates)	Oocyst Spike Dose (Mean ± SD)	Mean Percent Recoveries ± SD Sample Matrix			
		DI Water	Artificial Matrix	Raw Sewage	Primary Effluent
Direct (3)	95.9± 9.8	38.9 ± 5.2	60.7 ± 9.3		
IMS (3)	95.9± 9.8	47.6 ± 1.0	51.1 ± 5.9		
Direct (3)	95.3 ± 10.5		68.6 ± 6.6		20.9 ± 3.1
IMS (3)	95.3 ± 10.5		50.0 ± 15.4		36.7 ± 16.6
Direct (3)	476.5 ± 52.2		59.2 ± 9.0		17.0 ± 6.8
IMS (3)	476.5 ± 52.2		73.2 ± 6.3		43.8 ± 24.2
Direct (3)	118.4 ± 9.5			21.1 ± 3.4	
IMS (3)	118.4 ± 9.5			39.7 ± 4.3	
Direct (3)	558.9 ± 28.2			49.2 ± 5.9	
IMS (3)	558.9 ± 28.2			37.3 ± 9.7	

Note: No oocysts detected in unseeded samples analyzed.

In DI water, higher recoveries were obtained with samples subjected to IMS than with samples analyzed directly by microscopy. One reason for increased recoveries in IMS samples may be the concentration volume: IMS samples were concentrated to 10 mL whereas samples subjected to direct analysis were concentrated to 300 μL. Reducing the pellet to small volumes probably increased oocyst loss when samples were aspirated. In the artificial matrix, on two of three occasions, direct analysis revealed higher recoveries than samples subjected to IMS; however large standard deviations existed for these data, indicating high variability in oocyst counts. It was interesting to note that recoveries in the wastewater matrices were improved when compared to the DI water spiking studies, which was likely to be caused by oocyst adhesion to particulates, which probably improved their sedimentation during the centrifugation step.

We hypothesized that the inhibitory effects of large insoluble particles on IMS recovery efficiencies may be overcome by addition of a suitable buffer or chemical additive and, to this end, a chemical additive was sought. Oocyst recoveries in the artificial matrix exceeded 50% in the samples subjected to IMS. This prompted us to examine the constituents of the artificial matrix individually to determine possible inhibitory relief for the IMS. One constituent of the artificial matrix was kaolin or aluminum silicate (Sigma Chemical Co., St. Louis; catalog no. K7375). Initial trials with 0.1 g kaolin/250 mL concentrate of raw sewage indicated an improvement in oocyst recoveries from 39.5 ± 9.8% to 60.9 ± 4.9%. Recoveries increased to 72.6 ± 5.6 when 0.5 g of kaolin was used (data not shown). On the basis of these encouraging data, different amounts of kaolin (0.1–1.0 g) were examined for its effect on recovery of approximately 100 oocysts spiked into concentrates derived from 250 mL raw sewage samples. It was noted that the final sample concentrates post-IMS contained large amounts of kaolin particles and the effect of such particles on oocyst occlusion or loss from the well slides had to be excluded. This was achieved by performing an additional rinse of the IMS captured oocysts in the MPC-M before dissociating the organisms. These results confirmed that varying amounts of kaolin up to 0.75 g and performing an additional rinse did not adversely affect oocyst recoveries (data not shown).

In our initial trials no detergents were added to the samples before concentration and kaolin was not added to the IMS tube. It may have been possible to achieve higher recoveries using detergents and kaolin during the IMS procedure; however, we had not determined the effects that kaolin had on the performance of the IMS in recovering oocysts from raw sewage. From our trials to assess the effects of kaolin on oocyst recoveries it seemed plausible that increased volumes of raw sewage could be processed without compromising oocyst recoveries. Trials incorporating the addition of detergents to the sample during concentration and the use of kaolin to enhance the IMS performance were conducted in 500 mL samples. Results are presented in Table 2-2. As a reference point, four replicate volumes (500 mL) of reagent water were processed to determine the best-case scenario for recoveries using this method. Samples were concentrated using 500 mL conical centrifuge tubes. Mean oocyst recoveries were 51.1 ± 6.7% in reagent water. In raw sewage mean recoveries plummeted to 16.3 ± 9.6% in trial 1 and even further to 4.8 ± 1.4% in trial 2.

When 250 mL conical centrifuge tubes were used to concentrate the sample, a mean oocyst recovery of 72.9 ± 10.5% was achieved in reagent water (Table 2-3). Comparing these results with the oocyst recoveries using 500 mL conical tubes an increase of approximately 20% was observed. In 250 mL of raw sewage mean oocyst recoveries were 48.8 ± 18.0%, which dropped slightly to 42.9 ± 10.4 % in 500 mL and to 35.4 ± 17.6% in 1000 mL. Comparing mean recoveries in 500 mL of raw sewage an increase of greater than 32% was achieved when 250 mL conical centrifuge tubes were employed instead of 500 mL tubes.

Table 2-2. *Cryptosporidium* Oocyst Recovery in 500 mL Raw Sewage with Added Detergent, Concentrated By Centrifugation Using 500 mL Conical Centrifuge Tubes and Including Kaolin in the IMS Procedure.

Trial (no. of replicates)	Packed Pellet Volume (mL)	*C. parvum* Oocyst Spike Dose (Mean ± SD)	Percent Recovery (Mean ± SD)
1 (3)	0.25	106 ± 7	16.3 ± 9.6
2 (3)	1.0	105 ± 9	4.8± 1.4
DI (4)	Trace	105 ± 8	51.1± 6.7

No oocysts were detected in unseeded controls.

Table 2-3. Oocyst Recoveries in Raw Sewage with Added Detergent Using 250 mL Conical Centrifuge Tubes and Spike Dose Concentration of ~100 *C. parvum* Oocysts and Including Kaolin in the IMS Procedure.

Sample Volume (no. of replicates)	Range of Packed Pellet Volume (mL)	Percent Recovery (Mean ± SD)
250 mL (12)	0.4 – 1.0	48.8 ± 18.0
500 mL (12)	0.5 – 1.0	42.9 ± 10.4
1000 mL (12)	1.0 – 4.0	35.4 ± 17.6
250 mL (4, DI)	trace	72.9 ± 10.5

No oocysts were detected in unseeded controls.

These preliminary trials seemed to suggest that concentrating the wastewater with detergents and performing the IMS procedure in the presence of kaolin could result in reasonable recoveries. The detergent aids in the dissolution of fecal fats, oils, and grease that may be present in wastewater, allowing the surface epitopes of the oocyst to be exposed for facilitated IMS bead capture. Kaolin is added to sequester inert debris in the matrix, which then frees the oocysts to interact and bind with the IMS beads. These trials also suggest that using 250 mL conical centrifuge tubes for concentrating the wastewater samples contribute to higher recoveries. Further trials were conducted examining oocyst recoveries in raw sewage and primary effluent at 250–1000 mL samples. A total of 12 sampling events occurred with the raw sewage. Initial recovery trials were conducted with viable oocysts at spike concentrations of approximately 100 oocysts. The results from the raw sewage trials are presented in Table 2-4. Mean oocyst recoveries in 250 mL of raw sewage were 39.1–71.0% in trials 1 through 6. Mean oocyst recoveries were 38.1–47.2% and 21.4–48.8% in 500 mL and 1000 mL samples, respectively.

*Color*Seed became commercially available during the course of this research project. *Color*Seed is gamma-irradiated *Cryptosporidium* oocysts and *Giardia* cysts that have been stained with the fluorescent dye Texas Red, and counted using flow cytometry techniques to yield a suspension with a known number of organisms with a standard deviation of less than 2.5%. This product readily lends itself to serve as an internal positive control to determine overall method performance on a given sample where *Cryptosporidium, Giardia,* or both, are the target organisms. *Color*Seed organisms are detected during the microscopy phase of the analysis by using a standard green filter. *Color*Seed organisms are easily differentiated from indigenous oocysts because they fluoresce red using the green filter, whereas indigenous oocysts do not. Caused by the variability of the wastewater matrix an important step toward finalizing the methods for *Cryptosporidium* was incorporating *Color*Seed into the method to serve as an internal positive control.

Recovery of viable oocysts that were seeded into 250 mL raw sewage were 2.7–79.9% using the Dynal IMS kit with a mean recovery of 33.0 ± 21.1%. When the volume of raw sewage analyzed was increased to 500 mL, mean viable oocyst recovery dropped slightly to 31.8 ± 20.6%. Mean recovery of viable oocysts in 1000 mL raw sewage was 24.3 ± 19.9% using the Dynal IMS kit. Initial trials using *Color*Seed in 250 mL raw sewage, yielded mean recoveries of 39.9 ± 5.9% for viable oocysts compared to 35.7 ± 7.7 % for *Color*Seed oocysts (Table 2-4, trial 7). Further trials in raw sewage comparing viable versus *Color*Seed recoveries showed a marked decrease in oocysts recoveries for both types of oocysts (Table 2-4, trials 8 and 9). However, in each trial, the *Color*Seed recovery rate was similar to the rate of recovery of viable oocysts, suggesting that *Color*Seed could be a useful tool in determining overall method performance. In raw sewage trials using the Dynal IMS kit and *Color*Seed oocysts, mean *Color*Seed recoveries were 23.5 ± 13.8% for 250 mL of raw sewage, 18.1 ± 24.6 % for 500 mL, and 9.1 ± 13.5% for 1,000 mL. Mean recoveries of viable oocysts from raw sewage using the Dynal product were 22.0 ± 12.5% for 250 mL, 17.0 ± 22.0% for 500 mL, and 9.5 ± 11.3% in 1,000 mL [Table 2-4, trials 7–10 (Dynal data only) and 12].

During the course of this project, Aureon Biosystems released an IMS kit for the isolation of oocysts from environmental matrices. Validation trials conducted in source waters yielded results that met the acceptance criteria established for Method 1623 (U.S. EPA, 2001), and this kit was examined to see if it performed as well as the Dynal kit in wastewater. In head-to-head comparison trials in 250 mL raw sewage, viable oocysts recoveries were 17.7 ± 1.8% using Dynal kits and 15.7 ± 5.4% using Aureon kits (Table 2-4, trial 10). Mean recoveries of *Color*Seed in these trials were 37.1 ± 3.1% and 25.2 ± 3.2% for Dynal and Aureon, respectively. In this trial the *Color*Seed mean recoveries were greater than the viable mean recoveries by 20% for Dynal and 10% for Aureon. All trials conducted using the Aureon IMS kit yielded mean recoveries of viable oocysts of 17.3 ± 4.8% ($n = 6$), 7.2 ± 2.8% ($n = 3$), and 1.9 ± 1.2 % ($n = 3$) in 250 mL, 500 mL, and 1000 mL raw sewage, respectively (Table 2-4, trials 10 and 11), and compared to mean recoveries of *Color*Seed oocysts at 23.0 ± 4.4% in 250 mL, 6.8 ± 6.2 % in 500 mL and 2.0 ± 2.7 % in 1000 mL of raw sewage.

Statistical analyses were performed using SigmaStat™ software by Jandel Scientific. Statistical comparisons to assess the significance of differences in mean recovery rates were determined based on the null hypothesis. To determine equal sample means for recovery basic assumptions for parametric tests needed to be validated. These were normality and homogenecity of variance. Where assumptions were met, a one-way analysis of variance (ANOVA) was performed. Where ANOVA indicated inequality of sample means across the range of test conditions, further pairwise analyses were performed under a Student-Newman-Keuls multiple comparisons test. Where assumptions were not met application of nonparametric tests, Rank sum tests (Mann-Whitney) or (Kruskal Wallis ANOVA on ranks) were performed.

When 250 mL of seeded, raw sewage were examined across all experimental conditions, significant differences in the recovery rates of live oocysts or *Color*Seed using Dynal or Aureon were not observed ($P = 0.063$). However, when head-to-head trials in 250 mL raw sewage were analyzed, significant differences were noted in the recovery rates of *Color*Seed using Dynal versus Aureon ($P = 0.0004$). Significant differences were not observed on the 500 mL raw sewage data set ($P = 0.066$) again comparing viable oocyst recoveries and *Color*Seed recoveries using Dynal or Aureon IMS kits. Significant differences ($P = 0.027$) were revealed in the mean recoveries between these same test groups in 1000 mL raw sewage. Statistical tests of all raw sewage recovery data using the Dynal product failed to discern significant differences in recovery of viable oocysts as a function of sample volume ($P = 0.251$). For the Aureon product, however, viable oocyst recoveries from raw sewage were significantly different as a function of sample volume ($P = 0.0007$), with recoveries decreasing as a function of increasing volume.

Table 2-4. Recovery Rates of Live *Cryptosporidium* Oocysts and *Color*Seed in Raw Sewage.

Trial	IMS Kit (no. of replicates)	Volume Analyzed (L)	Packed Pellet Volume (mL)	Spike Dose of Live *C. parvum* Oocysts (Mean ± SD)	Percent Recovery of Live Oocysts (Mean ± SD)	Percent Recovery of *Color*Seed (Mean ± SD)
1	Dynal (3)	0.25	0.25	101.4 ± 9.2	71.0 ± 10.0	NA
	Dynal (3)	0.50	0.5		38.1 ± 9.4	NA
2	Dynal (3)	0.50	0.5	104.8 ± 8.5	47.2 ±12.6	NA
3	Dynal (3)	1.00	0.5	110.6 ± 9.8	48.8 ± 20.1	NA
4	Dynal (3)	0.50	0.5	101.0 ± 14.1	46.5 ± 15.6	NA
	Dynal (3)	1.00	1.0		37.6 ± 8.4	NA
5	Dynal (6)	0.25	1.0	101.1 ± 4.9	42.6 ±14.4	NA
	Dynal (3)	1.00	4.0		21.4 ± 18.4	NA
6	Dynal (3)	0.25	0.1	103.1 ± 4.8	39.1 ± 12.2	NA
	Dynal (3)	0.50	0.2		39.8 ± 3.5	NA
	Dynal (3)	1.00	0.25		33.9 ± 17.8	NA
7	Dynal (3)	0.25	0.2	90.3 ± 8.0	39.9 ± 5.9	35.7 ± 7.7 [1]
8	Dynal (3)	0.25	0.25	111.1 ± 12.3	5.7 ± 4.4	7.4 ± 1.5 [1]
	Dynal (3)	0.50	0.5		0.9 ± 0.9	1.0 ± 1.0 [1]
	Dynal (3)	1.0	1.0		1.2 ± 2.0	0 ± 0 [1]
9	Dynal (3)	0.25	0.2	105.6 ± 8.3	19.9 ± 9.1	9.2 ± 6.1 [2]
	Dynal (3)	0.50	0.4		4.7 ± 6.6	2.7 ± 3.8 [2]
	Dynal (3)	1.0	0.8		2.5 ± 1.1	0.7 ± 0.6 [2]
10	Dynal (3)	0.25	0.25	96.0 ± 7.4	17.7 ± 1.8	37.1 ± 3.1 [2]
	Aureon (3)	0.25	0.25		15.7 ± 5.4	25.2 ± 3.2 [2]
11	Aureon (3)	0.25	1.0	125.4 ± 14.2	18.9 ± 4.6	20.7 ± 4.8 [3]
	Aureon (3)	0.50	2.0		7.2 ± 2.8	6.8 ± 6.2 [3]
	Aureon (3)	1.0	4.0		1.9 ± 1.2	2.0 ± 2.7 [3]
12	Dynal (3)	0.25	0.2	113.1 ± 11.2	26.8 ± 2.1	27.9 ± 2.1 [3]
	Dynal (3)	0.5	0.4		45.4 ± 7.7	50.7 ± 3.6 [3]
	Dynal (3)	1.0	0.8		24.4 ± 2.0	26.5 ± 6.2 [3]

1. *Color*Seed spike dose concentration: 99 ± 1.2 *C. parvum* oocysts.
2. *Color*Seed spike dose concentration: 98 ± 1.7 *C. parvum* oocysts.
3. *Color*Seed spike dose concentration: 98 ± 1.4 *C. parvum* oocysts.
No oocysts were detected in unseeded samples.

Recovery data in primary effluent samples are presented in Table 2-5. Using the Dynal IMS kit, oocyst recoveries in primary effluent were slightly higher for both viable and *Color*Seed oocysts than in raw sewage. In 250 mL samples, viable oocyst recoveries were 10.9–107% and *Color*Seed oocyst recoveries were 24.5–52%. Mean recoveries in 250 mL samples were 42.7% ± 28.3% for viable oocysts and 36.2 ± 11.4% for *Color*Seed oocysts. The mean viable oocyst recovery in 500 mL was similar to that observed in 250 mL at 42.8 ± 31.5% whereas the mean *Color*Seed oocyst recovery was slightly higher than the 250 mL at 41.3 ± 14.7%. Mean oocyst recoveries in 1000 mL of primary effluent was 30.9 ± 23.6% and 43.2 ± 16.6% for viable oocysts and *Color*Seed, respectively. No recovery trials of oocysts from primary effluent using the Aureon IMS kit were performed.

In trial 2, the mean concentration in the unseeded samples was 73.3 ± 37.2 oocysts/L with oocyst concentrations were from 32/L in the 1000 mL subsample analyzed to 104/L in the 500 mL sample analyzed (52 oocysts/500 mL). Based on these analyses of the unseeded samples, the oocyst recoveries in this trial may be artificially high because of the high variability observed in the background oocyst concentration. ANOVA analysis revealed no significant differences in viable oocyst versus *Color*Seed mean recoveries ($P = 0.775$) regardless of the volume processed.

Table 2-5. A Comparison of Recoveries in Primary Effluent Using Live *Cryptosporidium* Oocysts and *Color*Seed.

Trial	Volume Analyzed (L) (no. of replicates)	Packed Pellet Volume (mL)	Turbidity (ntu)	Spike Dose of Live *C. parvum* oocysts (Mean ± SD)	Percent Recovery of Live Oocysts (Mean ± SD)	Percent Recovery of *Color*Seed (Mean ± SD)
1	0.25 (3)	0.1	41.3	95.3 ± 11.6	41.3 ± 17.0	NA
	0.50 (3)	0.15			15.0 ± 8.9	NA
	1.0 (3)	0.25			2.8 ± 2.2	NA
2	0.25 (3)	0.5	163	94.4 ± 10.8	90.0 ± 17.5 [1]	NA
	0.50 (3)	1.0			97.8 ± 15.9 [1]	NA
	1.0 (3)	2.0			66.0 ± 11.6 [1]	NA
3	0.25 (3)	0.1	77.7	112.6 ± 6.5	40.0 ± 2.3	NA
	0.50 (3)	0.2			32.6 ± 8.0	NA
	1.0 (3)	0.5			12.4 ± 0.9	NA
4	0.25 (3)	0.1	40.3	128.4 ±19.4	14.5 ± 6.3	31.6 ± 7.2 [2]
	0.50 (3)	0.2			23.4 ± 9.6	41.8 ± 22.1 [2]
	1.0 (3)	0.4			39.5 ± 6.3	56.8 ± 6.8 [2]
5	0.25 (3)	0.1	39.0	105.2 ± 5.6	27.6 ± 8.1	40.8 ± 14.4 [2]
	0.50 (3)	0.2			45.0± 5.7	40.8 ± 7.2 [2]
	1.0 (3)	0.4			33.9 ± 6.9	29.6 ± 9.1 [2]

1. Concentration of *C. parvum* oocysts in background samples were as follows: 21 oocysts in 250 mL, 52 oocysts in 500 mL and 32 oocysts in 1000 mL sample. No oocysts were detected in other unseeded samples.
2. *Color*Seed spike dose concentration: 98 ± 1.0 *C. parvum* oocysts.

Therefore, increasing the primary effluent volume analyzed to 1 L did not negatively impact oocyst recoveries and in fact, no discernable differences in mean oocyst recoveries were noted when a statistical analysis of the dataset was performed.

On the basis of the statistical analyses performed on each wastewater dataset, the Dynal IMS kit offers improved performance under most test conditions. Comparisons of Aureon and Dynal recoveries using viable oocysts and *Color*Seed oocysts in seeded raw sewage indicate no significant differences for the 250 and 500 mL samples. ANOVA revealed that mean recoveries using these two products were not the same for the 1000 mL samples, but it was not possible to discern pairwise differences under multiple comparisons analyses following the ANOVA on equal sample means. As volumes of raw sewage analyzed were increased from 250 mL to 1000 mL, no significant differences in viable oocyst recoveries were observed using Dynal. However, viable oocyst recoveries with Aureon were adversely affected as the sample size increased, with significant differences noted. The statistical analysis of each dataset for the raw sewage and primary effluent indicates that the number of trials performed to date is adequate in demonstrating significant differences in the following test parameters: (1) IMS product (Dynal vs. Aureon), 2) sample volume, and 3) oocyst source (*Color*Seed oocysts vs. viable oocysts). However, the recovery of *Color*Seed oocysts or seeded, viable oocysts may or may not correlate at all with the ability of the method to recover naturally occurring oocysts. The use of *Color*Seed is simply a tool to estimate the method performance in the matrix on a given day.

Possible differences in performance of each IMS kit could be caused by several factors including the size of the beads, the antibody conjugate or the IMS buffering systems. The Dynabeads are approximately 5.0 μm in diameter, whereas the Aureon beads are approximately

0.8 μm. When examining the beads under the microscope the Dynabeads have uniform size and shape, whereas the Aureon beads are not. Because of the uniform size and shape of the Dynabeads we suspect that a larger surface area comes in contact with the oocysts than with the Aureon beads, making the capture system more efficient. In addition, the antibodies conjugated to the beads for oocyst capture may be different for each IMS kit. These antibodies may target different surface antigens of the oocyst wall. Kuhn et al. (2002) noted mean oocyst recovery was increased 16.5% when the sample concentrates were adjusted to a neutral pH in the Dynal IMS procedure. Although not tested in this project, it is possible that the performance of the Aureon IMS kit did not adequately buffer the wastewater concentrates for optimal recovery

CHAPTER 3.0

RECOVERY OF *CRYPTOSPORIDIUM* OOCYSTS IN SECONDARY AND TERTIARY EFFLUENTS

3.1 Introduction

For sewage effluents, one method that has been described by the United Kingdom SCA requires filtration of 10–50 L of sewage effluent through a polypropylene cartridge filter at a flow rate of 1.5 L/min. The cartridge filters can then be processed according to the ICR method (U.S. EPA, 1996) or the SCA method in the United Kingdom (Anon., 1990). Oocysts are isolated from contaminating debris by using a cold sucrose solution of a specified density and then concentrated using centrifugation to a minimal volume. The second SCA method advocates screening of 5 L grab sample through a coarse (50–150 μm) filter, and concentrating by repeated centrifugation to 10 mL. The sample is subjected to sucrose density flotation and concentrated to a minimal volume (0.5–1.0 mL), depending on the levels of contaminating debris. Both these procedures were derived or suggested approaches rather than protocols developed after experimental validation. However, data have emerged indicating that 100 L filtered samples yield consistently lower numbers of oocysts than 2 L grab samples (Robertson et al., 2000; Bukhari et al., 1997). These investigations also indicated that for some filtered samples it was necessary to perform two or three sucrose density flotations before samples were in an acceptable state for microscopy.

Although sucrose density flotation is useful for clarifying oocysts from contaminating debris, there are several limitations to this technique. The number of oocysts present in the sample is likely to influence recovery efficiencies. For example, 1g bovine fecal samples seeded with 7 x 10^4 *C. parvum* oocysts yielded recovery efficiencies from 24–64.8% (Bukhari & Smith, 1995), whereas 2 L grab samples of sewage influents seeded with approximately 600 oocysts yielded recovery efficiencies between 6% and 21% (Robertson et al., 2000). When fecal and environmental samples were seeded with *Cryptosporidium* oocysts of known viability and purified by sucrose density flotation it was shown that selection for viable oocysts occurred (Bukhari & Smith, 1995, 1997). Because of these limitations with sucrose density flotation, an alternative, combined method, which used sucrose density flotation followed by diethyl-ether concentration of the sucrose pellet was developed for isolation of *Cryptosporidium* oocysts from sewage influent samples (Robertson et al., 2000; Bukhari et al., 1997). In addition, Bukhari and Smith (1995) reported that flotation procedures selectively concentrated viable oocysts and noted that oocyst recoveries were dependent upon population viability. Increased exposure to water concentrates has also been reported to reduce oocyst recoveries (Fricker, 1995), which was probably caused by increasing proportions of nonviable oocysts.

Several researchers have used spiral wound filters to determine oocyst concentrations in sewage effluents. In most of these studies recovery efficiencies were not assessed (Carraro et al.,

2000; Chauret et al., 1999; Dumoutier & Mandra, 1996; Gibson et al., 1998). Using Cuno filters for the collection of 10–50 L of settled raw influent followed by sucrose flotation, Medema and Schijven (2001) reported an oocyst recovery efficiency of 0.4%. Watt et al. (2002) compared oocyst recovery efficiencies using Cuno filters, electronegative-charged Filterite filters (nominal porosity of 0.45 µm) and 1MDS filters in 400 L of tertiary effluents. Oocysts were isolated from contaminating debris using Percoll-sucrose density gradient flotation and application of the resulting concentrate to PCTE membrane filters. Mean oocyst recovery efficiencies in tertiary effluents were 9.5 ± 3.7% using Cuno filters, 61.0 ± 21.0% using Filterite filters and 23.0 ± 3.4% using the 1MDS filters. One drawback using PCTE membrane filters for detection and enumeration of *C. parvum* oocysts is that internal structures of fluorescing objects cannot be examined using DIC optics to rule out cross-reacting species of algae or other biological particles. Our initial trials investigated oocyst recoveries in 50 L of secondary effluent using the spiral wound filter. Based on our experience in source water field spiking studies using spiral wound filters, we were aware that a majority of seeded oocysts passed through the filter when a flow rate of 4 L/min was applied (Clancy et al., 1997). The sample collection rate was reduced to 1 L/min and the oocysts were seeded in the first 10 L volume. The filter was washed according to ICR protocol with portions of the sample concentrate subjected to IMS for the specific isolation of *C. parvum* oocysts. The recovery rates for these trials were less than 20% (data not shown).

In 1997, the U.S. EPA began to develop a new method for protozoa analysis in source water, and in 1998, the first new method in over a decade was produced—U.S. EPA Method 1622: *Cryptosporidium* in Water by Filtration/IMS/FA (FA, fluorescent antibodies; U.S. EPA, 1997). Method performance was shown to be significantly better that the ICR method (U.S. EPA, 1996). Method improvements included: 1) a new filter, allowing complete capture with high and reproducible elution of oocysts from the filter, 2) introduction of a new technique—IMS—permitting specific capture of target organisms, and reducing false positives and background significantly, 3) an additional staining step to further aid in identification of *Cryptosporidium* oocysts, and 4) stringent quality assurance and quality control (QA/QC) as part of the method. Quickly following the commercial availability of IMS beads for the specific capture of *Giardia* cysts, Method 1623: *Giardia* and *Cryptosporidium* in Water by Filtration/IMS/FA was developed (U.S. EPA, 1999), and validated in collaborative trials (Connell et al., 2000).

Original validation studies for each method used the Pall Gelman Envirochek™ capsule filter. Since that time three other filters have been validated for use in Methods 1622/1623: the Corning CrypTest™ cartridge filter, the Pall Gelman Envirochek HV™ capsule, and the IDEXX Filta-Max™ system. Another filter that has been used to concentrate source water is the PCTE flat membrane. This membrane filter is available in 142- and 293-mm diameter formats and is available in a variety of pore sizes. The sampling strategies used for source waters are more appropriate for secondary effluents than raw sewage, because the various treatment processes that are employed in generating secondary effluent result in a considerable reduction in the microbial burden as well as contaminating debris from raw sewage. For these reasons it seemed appropriate to investigate the use of these filters for collection of 10 L of secondary effluent.

Standard Envirochek capsule filters were tested and found to be capable of collecting 10-L samples of secondary effluent (turbidity of 2.0 ntu). Pellet material generated from these 10-L samples was seeded with approximately 100 oocysts and recovered from the contaminating debris using IMS. Oocysts were recovered from the matrix at a mean recovery rate of 80.7 ± 9.9% (data not shown) indicating that the matrix did not inhibit the IMS performance. However, when 10 L samples were seeded with approximately 100 oocysts, filtered through the capsule filters, and the filter eluant subjected to IMS, mean oocyst recovery plummeted to 0.6 ± 0.5%. These data suggest the oocysts were embedded in the filter material and were not eluted using the standard Method 1622 elution protocol. Using an oocyst spike dose between 104 and 111 oocysts and 10 L samples, the prototype Envirocheck HV filters permitted overall oocyst recoveries less than 5%. Although various, more rigorous, elution protocols than the standard protocols employed in Methods 1622/1623 were examined, oocyst recoveries did not improve further (data not shown).

Using the Filta-Max filters, it was not always possible to collect the target 10 L samples of secondary effluent. The reasons for this are unclear and despite efforts to reduce compression in the filter module by reducing the number of pads from 60 to either 58 or 55, no improvement in flow rate was noted in some filters. The filtration equipment was examined for obvious obstruction; however, none were identified as the problem of poor flow rates occurred randomly. Using the standard elution protocol for the standard Filta-Max filters (60 pads) with the wash station and low spike doses into 10 L of secondary effluent, mean recoveries of 9.5% were achieved (data not shown). When reduced numbers of foam pads were used in each filter, oocyst recoveries increased to approximately 15% (data not shown). This occurred despite an increase in the sample turbidity from 3.5 ntu to 6.2 ntu. These data suggested that perhaps the particulates or chemical composition of the secondary effluent were responsible for trapping the oocysts within the filter matrix and the gentle washing procedure employed with the standard protocol (wash station) was inadequate to elute the oocysts. Using increased spike doses of oocysts (to facilitate direct enumeration of the filter eluant) comparative oocyst recovery studies were conducted in DI water and secondary effluent. Although the overall method (from sampling to IMS and detection) yielded recoveries of 60.1% in DI water, overall oocyst recoveries were 24.9% in 10 L of 4.4 ntu secondary effluent (data not shown).

This study also examined the suitability of 142 mm PCTE membrane filters for collection of 10 L of secondary effluent. Where secondary effluent turbidities ranged between 2.8 and 3.5 ntu, 10 L samples were filterable at flow rates up to 2 L/min. Eluting the filters by hand kneading and then concentration by centrifugation followed by IMS enabled recoveries of 44.9% (data not shown). To further optimize the elution step, the use of the stomacher was compared to hand kneading; however, the latter demonstrated more effective elution of spiked oocysts. Unfortunately, the disc membranes provide a limited surface area for sample filtration, making them prone to blockages when sample turbidities increase. For example, where sample turbidity was 4.4 ntu, only 5 L samples could be filtered through the membrane. Encouragingly, this did not impact the oocyst recoveries deleteriously and overall oocyst recoveries were approximately 65% (data not shown).

A number of different filters, which have demonstrated success in U.S. EPA Method 1622/1623, were examined for their suitability for collection and recovery of *Cryptosporidium* oocysts from secondary effluent. Filters such as the Gelman Envirocheck and prototype Envirocheck HV performed poorly, with marginally better performance observed when using yarn wound filters followed by the Filta-Max filters. Both the CrypTest and 142 mm diameter membrane filters demonstrated overall oocyst recoveries approximately of approximately 50%.

Unfortunately, the sensitivity of the 142 mm membrane filters to sample turbidity variations means it is unlikely that they will be successful for handling a wide variety of secondary effluents. In contrast, the pleated nature of the CrypTest filters, which helps to increase their surface area significantly, when combined with the improved elution procedure for recovery of entrapped oocysts, seems to indicate their potential suitability for secondary effluents. Trials conducted to determine oocyst recoveries in secondary effluent are presented below.

When Pall Corp. released a new design of the Envirochek HV filter it was included in this study to examine its usefulness in recovering *C. parvum* oocysts in secondary effluent. Trials performed in the past using a prototype HV filter yielded oocyst recoveries less than 5%. With the new HV filter using the standard elution protocol, oocyst recoveries were 21.3–49.2% with a mean recovery of 32.5 ± 12.5%. These recoveries were a significant improvement over those previously obtained with the prototype. In drinking water research, also conducted in this laboratory, we had demonstrated when using the HV filter that oocyst recoveries were dramatically increased by incorporating a pretreatment with 5% sodium hexametaphosphate solution (NaHMP) (McCuin et al., 2001b). These results prompted us to conduct trials using the HV filter and to include a pre-elution step with NaHMP.

3.2 Materials and Methods

3.2.1 *Cryptosporidium* Oocyst Spike Dose Source and Enumeration Procedures

Same as described in Chapter 2.0, Section 2.2.1.

3.2.2 Turbidity Measurements

Turbidity of the secondary and tertiary effluents was determined using a HACH Model 2100P turbidimeter capable of measuring turbidity levels between 0 and 1000 ntu. Calibration of the instrument was performed each day of use with standards prepared by the manufacturer. No dilution of the matrix was required for determination of sample turbidity.

3.2.3 Epifluorescence Microscopy

Performed as described in Chapter 2.0 Section 2.2.5.

3.2.4 *Cryptosporidium* Oocysts Recoveries Using CrypTest Filters

3.2.4.1 CrypTest Trials Using Standard Elution Procedure

Three CrypTest filters were prespiked with approximately 100 *C. parvum* oocysts and then used to collect secondary effluent at a flow rate of 2 L/min. One unspiked filter was also used to collect secondary effluent and analyzed to determine presence of background numbers of oocysts. Secondary effluents used for assessing recoveries from CrypTest filters had a turbidity of 6.5 ntu. This increased turbidity seemed to limit sample volumes through each filter to 6.0–7.5 L. Each CrypTest filter was backwashed to dislodge captured oocysts from the filter material. This was achieved by using a peristaltic pump to introduce 80% PBS/20% ICR buffer (80/20) at a flow rate of 200 mL/min into the housing through the outlet port until it was full. The housing was placed in sonicator bath for 2 min and the backwash solution was poured out of the housing via the inlet port. This backwashing procedure was repeated and after the eluant was decanted from the housing, compressed air was used to remove any remaining fluid in the filter. The excess fluid was also added to the backwash eluant. The pooled backwash was transferred to

250-mL centrifuge tubes and concentrated (1500 x *g*; 15 min). The supernatant was aspirated to 10 mL, the pellet vortexed and transferred to a Leighton tube for oocyst isolation by IMS as described in Method 1623. For secondary effluents it was not necessary to include kaolin in the IMS procedure. Recovered oocysts were enumerated by using epifluorescence microscopy.

3.2.4.2 Modified Backwash Procedures for CrypTest Filters

Triplicate 10 L samples of secondary effluent were seeded with approximately 100 *C. parvum* oocysts and filtered through a CrypTest filter at a flow rate of 2 L/min. Each carboy was rinsed with a 1 L PBST and passed through the filter. Each filter was then backwashed with PBST three times using the protocol described above. The backwash and excess fluid was pooled in a 1 L bottle. The pooled backwash was concentrated by centrifugation (1500 x *g*; 15 min), the supernatant was aspirated to 10 mL and the pellet was transferred to a Leighton tube containing IMS buffers. The oocysts were recovered from the sample using IMS and enumerated using immunofluorescence microscopy. This procedure was repeated on a different day with three replicate 10 L samples of secondary effluent and duplicate 10 L samples of DI water.

3.2.4.3 Incorporating NaHMP Into the Elution Procedure for CrypTest Filters

We were interested in determining if using NaHMP would increase oocysts recoveries as observed with the HV filter. Secondary effluent samples (10 L) were seeded and filtered through CrypTest filters as previously described. NaHMP solution was introduced into the sample using the backwash technique. The filter and housing were placed on the wrist shaker and shaken at full speed for 5 min. The NaHMP was pulled through the filter using a vacuum pump. The filter was rinsed with DI water to remove excess NaHMP. Following the DI water rinse, *C. parvum* oocysts were eluted from the filters as previously described. The resulting filter wash was concentrated by centrifugation (1500 x *g*; 15 min) and the supernatant aspirated. Sample concentrates were subjected to IMS and recovered oocysts were enumerated using epifluorescence microscopy.

3.2.5 *Cryptosporidium* Oocyst Recoveries Using Envirochek HV Filters

3.2.5.1 Carboy Spikes

Envirochek HV filters (Pall Corp., Ann Arbor, MI) were used to capture and retain oocysts in secondary effluents. During method development trials, spike doses of approximately 100 viable *C. parvum* oocysts were seeded into 10 L of secondary effluent. The seeded carboy, containing a magnetic stir bar was placed on a magnetic stir plate. The stir plate was turned on and allowed to mix thoroughly before filtration. The filter was placed at the outlet port of the carboy. Following the filter was a centrifugal pump equipped with an adjustable flow control valve (See Figure 3-1).

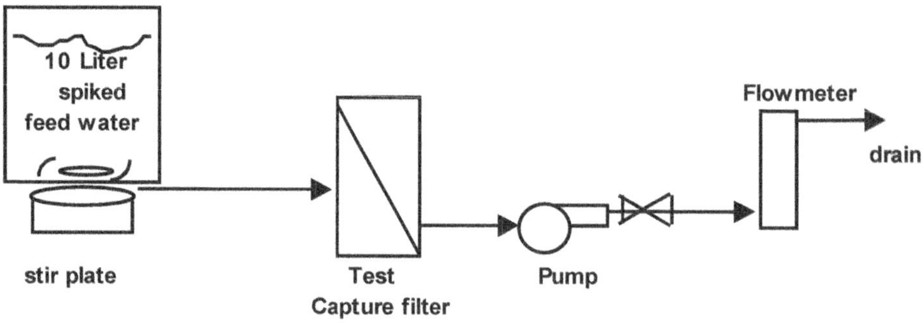

Figure 3-1. Carboy Spiking Setup for Secondary Effluents.

After the entire 10 L of secondary effluent had passed through the filter at a flow rate of 2 L · min^{-1}, the carboy was rinsed with 1 L PBST to remove residuals from the sides of the carboy. This volume was passed through the filter. A pre-elution step was performed by adding a 5% (w/v) solution of NaHMP through the inlet port with a volume sufficient to cover the pleated membranes in the capsule. The capsule was placed on the wrist shaker (Lab-Line Instruments, Inc., Melrose Park, IL, Model no. 3589) and shaken at maximum speed (750 rpm) for 5 min. The solution was pulled through the filter, so that dispersed material could exit through the filter pores, but oocysts and other intact particles greater than 1 µm would remain in the capsule. This action was followed by a DI water rinse, which was poured into the inlet of the capsule and allowed to drain through the filter membrane, continuing the purge of dissolved or submicron material. These preliminary elution steps were followed by the standard Method 1622 procedure (U.S. EPA, April 2001). Briefly, Laureth-12 buffer was added to the top of the pleats of the capsule. The filter was placed on the wrist shaker with the Luer-lock vent at the 12 o'clock position. The filter was then shaken at maximum speed for 5 min. The filter wash was decanted to a 250 mL conical centrifuge tube. Fresh Laureth-12 buffer was added above the pleats of the capsule and placed on the wrist shaker with the Luer-lock at the 4 o'clock position. The filter was then shaken for 5 min at full speed. Without decanting the filter wash, the filter was rotated so that the Luer-lock was at the 8 o'clock position and shaken for another 5 min. At this point, the filter wash was decanted to the 250 mL conical tube containing the initial sample eluant and was concentrated by centrifugation (1500 x *g*; 15 min). The supernatant was aspirated to 5–8 mL and then quantitatively transferred to a Leighton tube containing IMS buffers. *C. parvum* oocysts in the concentrates were isolated using either the Dynal or Aureon IMS kits as described in Chapter 2.0, Section 2.2.3 and 2.2.4, and recovered organisms were enumerated using epifluorescence microscopy.

In later trials *Color*Seed was included to compare recoveries with viable oocysts. Briefly, viable oocysts (~100) were seeded to 10 L volume of secondary effluent as well as the contents of one vial of *Color*Seed. The transfer protocol was described in Chapter 2.0, Section 2.2.1.2. The sample was allowed to mix and was then filtered through an HV filter at a flow rate of 2 L min^{-1} as described above. The filter elution procedure was performed as described in the previous paragraph. Oocysts were recovered from the resulting concentrates using Dynal or Aureon IMS procedures described in Chapter 2.0.

3.2.5.2 Field Collection Procedures for Tertiary Effluents Using Envirochek HV Filters

For tertiary effluents, viable oocysts were seeded directly into the capsule before sample filtration. The rationale for performing the seeding trials in this manner was the large sample volumes assayed for this matrix. It was impossible to transport 400 L of tertiary effluent—required for each seeding trial—to the laboratory. Adding the seed dose first offered the advantage of knowing that the entire dose was in the filter regardless of the volume filtered. In addition, by applying the seed dose first, a worst-case scenario was created where the seed dose was underneath the sample. For each matrix tested an unseeded sample was also analyzed to determine background concentrations of the target organism. When *Color*Seed was incorporated into the assay the contents of the *Color*Seed vial was decanted to the inlet port of the HV filter using the procedure prescribed by the manufacturer. Tertiary effluent sample collected on HV filters were processed in the sample manner as the secondary effluents.

3.3 Results and Discussion

Trials using CrypTest filters for capturing and recovering *C. parvum* oocysts in secondary effluent are presented in Table 3-1. The standard elution protocol using two backwashes yielded poor oocyst recoveries (Table 3-1, trial 1). However, when the backwashing procedure was modified to use a total of three washes and PBST substituted for the original eluting solution, mean oocyst recoveries in 10 L of secondary effluent improved significantly, between 43.7 and 55.7% (Table 3-1, trials 2 and 3b). Recoveries in DI using the modified procedure resulted in a mean oocyst recovery of 56.8%. It is possible that the low recoveries noted in trial 1 may have been caused by undetermined matrix interferences.

Table 3-1. Recoveries of *C. parvum* Oocysts Using CrypTest Filters.

Trial (*n*)	Sample Matrix	Turbidity (ntu)	Volume Filtered (L)	Spike Dose(Mean ± SD)	Packed Pellet Volume (mL)	Mean Recoveries (%)	Comments
1 (3)	SE	6.5	6.0–7.5	111.7± 7.9	0.5	6.0 ± 4.5	2 Backwashes with 80/20
2 (1)	SE	4.9	10	83.2 ± 10.2	0.5	50.5	3 Backwashes with PBST
3a (2)	DI	—	10	91.5 ± 11.7	Trace	56.8 ± 10.8	3 Backwashes with PBST
3b (3)	SE	4.9	10	91.5 ± 11.7	0.5	51.7 ± 6.9	3 Backwashes with PBST

SE, secondary effluent.
DI, deionized water.

In these trials the data presented seem to suggest that the CrypTest filter showed promise for both retention and subsequent recovery of *C. parvum* oocysts seeded into secondary effluent. These trials also indicate that employing the standard elution protocol (two backwashes with 80/20 buffer) yielded low recoveries; however, when the elution buffer was changed to PBST and three backwashes were employed, oocyst recoveries increased to approximately 50%. Trials were conducted to determine whether it was the increase in the number of backwashes, the change in elution buffer from 80/20 to PBST, or both, were responsible for improved oocyst recoveries. Direct analysis on each eluant from the CrypTest filter backwashes indicated similar recoveries for both elution buffers (data not shown). For these trials, the turbidity of the secondary effluent samples was 1.7 ntu and it was highly likely that low total percentage recoveries for the three backwashes were becaus the organisms were occluded during their direct analysis (see Table 3-2, trials 1 and 2). This was the case when oocysts from each backwash were individually recovered using IMS and resulted in improvements in oocyst recoveries compared to direct analysis (Table 3-2, trials 3 and 4). Furthermore, comparison of both elution buffers (PBST and 80/20) also indicated that no differences existed between these two buffers with respect to oocyst elution from CrypTest filters.

Table 3-2. Comparison of Two Elution Buffers on Oocyst Recoveries from CrypTest Filters.

Trial (*n*)	Turbidity of Secondary Effluent (ntu)	Elution Buffer	Mean Percent Oocyst Recovery ± SD by Direct Examination	Mean Percent Oocyst Recovery± SD by IMS
1(2)	1.7	PBST	15.3 ± 2.4	-
(2)		80/20	14.8 ±0.8	-
2 (2)		PBST	25.3 ± 17.1	-
(2)		80/20	15.7 ± 4.4	-
3 (2)	1.2	PBST	-	29.2 ± 1.5
(2)		80/20	-	31.3 ± 0
4 (2)		PBST	-	28.3 ± 4.8
(2)		80/20	-	20.4 ± 6.4

A concerning observation was that overall oocyst recovery declined from approximately 50% to less than 30%. Trials using control spikes in DI water demonstrated recoveries of 78.4 ± 17.3% from CrypTest filters (data not shown), which excluded questions regarding filter membrane integrity and suggested the likelihood that the reduction in oocyst recoveries was associated with variability in the secondary effluent matrix that might be independent of sample turbidity. For example, approximately 50% recoveries were obtained with secondary effluent with turbidity of 4.9 ntu, whereas samples with turbidities of 1.2–1.7 ntu yielded recoveries of approximately 30%. It is possible that treatment chemicals used in the coagulation process or the composition of the waste being treated needs to be considered when trying to recover oocysts from complex and variable matrices. It is possible that polymers used in the treatment process help to enmesh organisms within their matrix, preventing their elution with gentle procedures such as backwashing. In contrast, adequate shearing forces may be generated during the more rigorous elution procedures to break these polymer chains thereby enabling improved oocyst elution. These data also suggest that the CrypTest filter may not the best option for retention and elution of *C. parvum* oocysts in secondary effluent because of matrix properties of the secondary effluent. It seems that their pleated nature can be a double edged sword in that it helps to increase the surface area to improve the filters capacity for sample collection, but the pleats may also complement the effect of flocs and polymers that may be present in samples with respect to organism adhesion within the filter matrix. We were interested in determining if including a pre-elution step with NaHMP would increase oocysts recoveries. During the elution step NaHMP solution was introduced into the sample using the backwash technique as previously described. The filter and housing were placed on the wrist shaker and shaken at full speed for 5 min. The NaHMP was pulled through the filter using a vacuum pump. The filter was then rinsed with DI to remove excess NaHMP. Following the DI rinse *C. parvum* oocysts were eluted, concentrated and isolated from the debris as previously described. Oocyst recoveries were 29.1–51.2% with a mean oocyst recovery of 37.9 ± 11.7 % (data not shown). It is not clear from these limited trials if the NaHMP offers any advantage to recovering oocysts from the filter.

In drinking water research, we demonstrated when using the HV filter that oocyst recoveries in finished water are dramatically increased by incorporating a pretreatment with 5% NaHMP. NaHMP has been used for many years to reduce scaling (Fireman & Reitemeier, 1944), and is the main ingredient used in household products to aid cleansing in hard water conditions. In drinking water applications, it is used as a scale inhibitor in membrane pre-treatment (Schock, 1990; Kasper, 1993) and in lime softening processes. Because treatment chemicals used to treat wastewater are similar to those used in drinking water it seemed prudent to conduct experiments to determine if oocyst recoveries could be increased by using this pretreatment. In preliminary trials using NaHMP a mean oocyst recovery of 32.8 ± 15.5% did not exceed the recovery rate of a single filter processed without the chemical (49.2%) (see Table 3-3). However, in a subsequent trial the mean oocyst recovery with NaHMP was almost double that of those analyzed without, 46.8% and 26.9%, respectively (see Table 3-3).

Table 3-3. *Cryptosporidium* Oocyst Recoveries in 10 L Secondary Effluent Using Pall Envirochek HV Filters.

Trial (*n*)	Turbidity (ntu)	Filter Elution	*C. parvum* Oocyst Mean Spike Dose	Percent Recovery (Mean ± SD)
1 (1)	2.28	Without NaHMP	91.0 ± 7.6	49.2
(2)		**With NaHMP**		**32.8 ±15.5**
2 (3)	3.21	Without Na HMP	112.6 ± 6.5	26.9 ±6.9
(3)		**With NaHMP**		**46.8 ± 23.2**
Mean Recovery without NaHMP: 32.5 ±12.5			**Mean Recovery with NaHMP: 41.2 ±19.6**	

Recovery data for secondary and tertiary effluents collected using Envirochek HV filters are presented in Table 3-4. All samples were eluted incorporating 5% NaHMP as a pretreatment buffer for dispersing the material collected on the membrane surface followed by a DI water rinse. Following this pretreatment, the retained oocysts were eluted from the filter and isolated from the interfering debris using IMS as described in Method 1622. Differences in viable oocyst recoveries using the two IMS kits were most prevalent in the secondary and tertiary effluents. Mean viable oocyst recovery in secondary effluent was 53.0 ± 19.2 % with recoveries from 21.9–75.2% using the Dynal IMS kit. Viable oocyst recoveries using the Aureon IMS kit were 8.8–10.4% ($n = 3$) with a mean of 9.3 ± 0.9%. Mean *Color*Seed oocyst recoveries in secondary effluent were 30.4 ±17.5 % using Dynal IMS kit compared to 3.4 ± 0.6% using the Aureon kit. ANOVA revealed significant differences ($P = 0.0003$) in this comparison of recovery rates. Multiple comparisons analysis demonstrated the most significant differences in live oocyst recoveries using Dynal vs. Aureon, with a difference of the mean recoveries of 43.6% as well as recovery rates for *Color*Seed versus live oocysts using Dynal.

A mean recovery of 67.8 ± 4.4% was achieved using Dynal compared to 41.2 ± 13.8% with Aureon in isolating viable oocysts from tertiary effluent. Significant differences were noted ($P = 0.0014$) in the recovery of viable oocysts using the two products with differences in mean recoveries of 26.6%. *Color*Seed was recovered at a rate of 48.6 ± 12.3% and 28.1 ± 13.5% using Dynal and Aureon, respectively with no significant differences noted.

The results of these trials suggest that the performance of the Dynal IMS product is superior to the Aureon kit in recovering *C. parvum* oocysts from secondary and tertiary effluents. The viable oocyst recoveries obtained with the HV filters using the Dynal IMS kit are significantly higher than the recoveries obtained with the CrypTest filters most likely because of

Table 3-4. Oocyst Recoveries in 10 L Secondary and Tertiary Effluents Using Live *Cryptosporidium* and *Color*Seed.

Matrix	IMS Kit (no. of replicates)	Volume Analyzed (L)	Packed Pellet Volume (mL)	Turbidity (ntu)	Spike Dose of Live *C. parvum* oocysts (Mean ± SD)	Percent Recovery of Viable Oocysts (Mean ± SD)	Percent Recovery *Color*Seed (Mean ± SD)
Secondary Effluent	Dynal (2)	10	0.5	2.28	91.0 ± 7.6	32.8 ±15.5	NA
	Dynal (3)	10	0.2	1.99	113.0 ± 16.3	54.3 ± 14.3	18.7 ±16.5 [2]
	Dynal (3)	10	0.1 – 0.7	3.21	112.6 ± 6.5	46.8 ± 23.1	NA
	Aureon (3)	10	0.1	2.90	125.4 ± 14.2	9.3 ± 0.9	3.4 ± 0.6 [3]
	Dynal (3)	10	0.1	2.34	113.1 ± 11.2	71.3 ± 3.4	42.2 ± 8.9 [3]
Tertiary Effluent	Aureon (3)	75.7	0.05	- [1]	125.4 ± 14.2	57.7 ± 5.6	44.6 ± 8.7 [4]
	Aureon (3)	62.0	0.05	- [1]	115.2 ± 5.9	38.2 ± 3.5	21.1 ± 5.0 [4]
	Aureon (3)	52.0	0.05	- [1]	128.4 ± 19.4	27.8 ± 5.3	18.7 ± 4.2 [4]
	Dynal (3)	39.2	0.05	- [1]	113.1 ± 11.2	67.8 ± 4.4	48.6 ± 12.3 [5]

1. Turbidity of tertiary effluent was approximately 2 ntu.
2. *Color*Seed spike dose concentration: 99 ± 1.2 *C. parvum* oocysts.
3. *Color*Seed spike dose concentration: 98 ± 1.7 *C. parvum* oocysts.
4. *Color*Seed spike dose concentration: 98 ± 1.0 *C. parvum* oocysts.
5. *Color*Seed spike dose concentration: 98 ± 1.4 *C. parvum* oocysts.

No oocysts detected in unseeded samples.

the vigorous elution technique employed for the HV filters. Although fewer trials were conducted on the secondary and tertiary effluents when compared to the number conducted on raw influent and primary effluents, statistical analysis of these data sets indicates that an adequate number of trials were conducted to determine differences in the parameters analyzed (e.g., IMS kit, viable vs *Color*Seed recovery rates).

In a study comparing *Cryptosporidium* oocysts and *Giardia* cyst recoveries of *Easy*Seed and *Color*Seed, in source and finished waters, significantly lower mean recoveries were noted with *Color*Seed (Warnecke et al., 2003), although the reported differences were less than 5%. In 494 test samples of source and finished waters, mean *Easy*Seed oocyst recoveries were 63 ± 19% and mean *Color*Seed oocyst recoveries were 60 ± 18%. *Giardia* cyst recoveries reported for these samples 78 ± 14% for *Easy*Seed and 74 ± 15% for *Color*Seed. Warnecke et al. proposed that the lower *Color*Seed recoveries were because of the increased hydrophobicity of the Texas Red-labeled organisms. The treated organisms may have a greater tendency to attach to surfaces and are thereby lost during sample processing. Francy et al. (2004) reported no significant differences in recovery results obtained using flow-counted viable oocysts and *Color*Seed oocysts in stream samples. Oocyst recoveries were 2.0–61% and 3.0–59% for viable and *Color*Seed oocysts, respectively, in 30 environmental waters tested.

In the present study, lower *Color*Seed oocysts recoveries were obtained in the secondary and tertiary effluents as compared to the recovery of seeded viable oocysts in these matrices. However, these differences were, in some cases far greater than 5%. It seems unlikely that a majority of *Color*Seed oocysts were lost during sample processing because of their hydrophobic nature. We continue to suggest that the recovery of *Color*Seed oocysts or seeded viable oocysts may or may not correlate with the ability of the method to recover indigenous oocysts. The use of *Color*Seed as an internal positive control is simply a tool to estimate method performance in a matrix on a given day

CHAPTER 4.0

RECOVERY OF *CRYPTOSPORIDIUM* OOCYSTS IN BIOSOLIDS

4.1 Introduction

A variety of techniques for isolation of *C. parvum* oocysts from fecal material have been investigated. These include zinc sulfate, sucrose density, saturated salt solution flotation, Percoll-sucrose density gradient flotation, and formol-ether concentration. Bukhari and Smith (1995) reported oocyst recoveries of 46–75% when bovine feces (1 g) were seeded with high oocyst doses and subjected to water-ether concentration. In this same study, sucrose (sp gr 1.1) and zinc sulfate (sp gr 1.18) flotation techniques yielded oocyst recoveries were 24–65% and 22– 41%, respectively. In the United States, the flammable nature of diethyl ether and its potential carcinogenic properties has led to contraindication of its use. As the goal of this research has been to develop user-friendly procedures, this approach was abandoned and alternative procedures, based on specific oocyst clarification with IMS, were targeted. Rimhanen-Finne et al. (2001) isolated *C. parvum* oocysts from 1 mL sewage sludge by diluting to 10 mL with DI water and then subjecting the sample to IMS followed by PCR. Although IMS recovery rates of seeded organisms were not assessed, the sensitivity of the method was reported to be 625 cysts and oocysts/mL of sludge. Chauret et al. (1999) detected oocysts in mixed liquor, decant liquor, and sludge cake when these samples were resuspended in 1% (w/v) Tween 80 and oocysts were isolated from contaminating debris using the isolation and detection techniques described in the ICR method. Recovery rates for this method were not assessed. Massanet-Nicolau (2003) reported oocyst recoveries between 40–60% when biosolids (primary settled sewage sludge, digested sludge, and bovine manure) were subjected to sedimentation and IMS as compared to recovery rates less than 10% using sucrose flotation and IMS. However, Bukhari and Smith (1995) reported that flotation isolation technique selectively concentrates viable oocysts.

Recovery trials were conducted to improve or develop methods for recovering *Cryptosporidium* oocysts from various sludge matrices using a variety of separation and isolation techniques such as IMS, coarse filtration through muslin, density gradient flotation, and slow-speed centrifugation. The "White House method" (named for photo of a bulldozer digging up the White House lawn adorning the cover of the report) for recovering *Ascaris* ova from sludge was also investigated as a potential method for recovering oocysts in biosolids (U.S. EPA, 1992).

4.2 Materials and Methods

4.2.1 *Cryptosporidium* Oocyst Spike Dose Source and Enumeration

Cryptosporidium spike dose material was prepared and enumerated as described in Chapter 2.0, Section 2.2.1.

4.2.2 Description of Biosolids Samples

Digester sludge from the St. Albans Wastewater Treatment Facility was collected in 1-L polypropylene bottles. The WWTF personnel calculated the percent solids to be 9.5%, determined by desiccating a subsample of sludge in an oven at 103–105°C until a constant weight was obtained. The weight of the dried sample (minus the weight of the empty weighing dish) was divided by the weight of the wet sample (minus the weight of the empty dish) and multiplied by 100 to calculate percent solids. In addition, two types of sludge were received from Allegheny County Sanitary Authority (ALCOSAN) in Pittsburgh, Pa. One type was activated dewatered sludge with a solids concentration of 18–22%; the other was lime-stabilized sludge with 25–27% solids.

4.2.3 Methods for Recovery of *Cryptosporidium* Oocysts in Biosolids

4.2.3.1 Recovery of Oocysts in 5 g of St. Albans Sludge Using Dynal anti-*Cryptosporidium* IMS Kit

Six replicate 5-g subsamples of St. Albans sludge were weighed into tared individual 30 mL glass beakers and diluted in 4–5 mL DI water. Each sample was mixed gently to create a uniform suspension and then decanted into a Leighton tube. DI (2–3 mL) was added to each beaker to rinse the residual sludge suspension. One milliliter each of IMS buffer, SL-A, and SL-B was added to each beaker to rinse residual debris and transferred to the respective Leighton tube. A predetermined number of *C. parvum* oocysts (~100) were added to each resuspended sludge sample. Each sample tube was rotated gently several times to mix the sample. One 5-g subsample of sludge was unseeded to determine the background levels of oocysts. Anti-*Cryptosporidium* IMS beads (100 µL) were added to each Leighton tube. The samples were allowed to rotate for 1 h and the bead-oocyst complexes were isolated from the supernatant by application of magnetic separation. The bead-oocyst complexes were resuspended in 10 mL PBS to perform one additional rinse. Following this, the IMS procedure and detection using immunofluorescence microscopy were performed as previously described.

Trials were also extended to evaluate the effect of the overall procedure by spiking the rehydrated sludge in the 30-mL beaker with *C. parvum* oocysts and then transferring the spiked suspensions into Leighton tubes.

4.2.3.2 A Comparison of Oocyst Recoveries in 5 g St. Albans Sludge With and Without Kaolin

To determine whether kaolin exerted a beneficial effect on oocyst recoveries from sludge samples, 11 replicate 5 g subsamples of sludge were prepared for analysis as described above. Five of the seeded replicates and one unseeded replicate were processed using the protocol described above, whereas kaolin (0.75 g) was added to each of the five remaining seeded replicates. All samples were subjected to the IMS procedure and recovered oocysts were enumerated using immunofluorescence microscopy.

4.2.3.3 Oocyst Recoveries in 10 g and 15 g Samples Using IMS

To determine if larger amounts of sludge could be analyzed, trials were performed by suspending 10 g and 15 g of St. Albans sludge in 20 mL and 30 mL of PBST, respectively. Each suspension was seeded with approximately 100 oocysts, thoroughly mixed, and then divided between two Leighton tubes. Seeded oocysts were recovered using IMS protocols with the following exceptions. Instead of adding 100 µL of beads to each Leighton tube, 50 µL of beads were used. The bead-oocyst complex was isolated using IMS, however when the complex was suspended in 1 mL SL-A buffer, both suspensions were pooled into a single Eppendorf tube for each sample.

4.2.3.4 Recovery of *C. parvum* Oocysts from Sludge Using IMS and a Flat Plate Magnet

Six 5 g subsamples of sludge from St. Albans WWTF and six 5 g subsamples from ALCOSAN lime-stabilized sludge were weighed into individual 50 mL glass beakers, diluted with DI water to a total sample volume of 20 mL and five subsamples from each site were seeded with approximately 100 *C. parvum* oocysts. For each site, the sixth biosolids replicate was analyzed for the presence of indigenous oocysts present. After addition of IMS buffers (1 mL each of SL-A and SL-B) and 100 µL IMS beads to each sample, each beaker was covered with parafilm and placed in a shaking incubator (Lab-line Orbit Environ Shaker) at 50% speed for 1 h. Following this, contents from each beaker were transferred into a 100 x 15-mm polystyrene Petri dish. An additional 5 mL DI water was used to rinse the debris remaining in the beaker and this was pooled with its respective sample in the Petri dish. The Petri dish was placed on a plastic disk equipped with a large flat magnet in the middle of the plate, which was placed on an orbital mixer (Boekel Orbitron Rotator model 260250) and allowed to mix for 10 min at maximum speed (25 rpm). The disk containing the Petri dish was removed carefully from the orbital mixer and contents of the Petri dish were poured into a waste container. Excess debris remaining in the Petri dish was rinsed by adding 10 mL DI water, and discarding the rinse. This rinsing step was repeated. The bead-oocyst complex remaining in the Petri dish, was suspended in 1 mL SL-A buffer, transferred to a 1.5-mL Eppendorf tube and the tube was placed in the MPC-M to separate target organisms from interfering debris. To effectively increase bead-oocyst recovery from the Petri dish a second 1 mL rinse was employed with PBST. After separation with the MPC-M, the supernatant was removed from the Eppendorf tube and the second 1 mL rinse (PBST) was added to the same Eppendorf tube. The oocyst-bead complex was separated from the interfering debris and the standard IMS procedure was followed. Duplicate 5-g samples of ALCOSAN dewatered sludge were also seeded with approximately 100 *C. parvum* oocysts and processed as described.

4.2.3.5 Recovery of Oocysts in Supernatant and Settled Sludge Suspension Using Coarse Filtration, Density Gradient Flotation, and IMS

Lime-stabilized sludge (100 g) from ALCOSAN was suspended in 1 L PBS and 10 mL Tween 20. *C. parvum* oocysts ($\sim 10^3$) were added, the sample was vigorously shaken for 30 min in the shaking platform incubator. The bottle was removed from the incubator and allowed to stand for 20 min. The supernatant was transferred to 250-mL conical tubes, concentrated by centrifugation (1500 x *g*; 10 min) and the supernatant was aspirated to above the pellet material. The pellet material was resuspended in PBST to 10 mL and transferred to a Leighton tube to recover oocysts by IMS. The remaining sludge suspension was passed sequentially through 149-mm and 70-mm nylon mesh, rinsing particles retained on the mesh with copious amounts of PBST. The filtrate was concentrated by centrifugation in 250-mL conical tubes. The supernatant was aspirated and the pellet material was resuspended to 125 mL with PBST. A 15 mL subsample was processed by IMS and recovered oocysts were enumerated using immunofluorescence microscopy. Another 40 mL subsample was divided between two 50-mL conical tubes. Each sample was subjected to density gradient flotation using Percoll-sucrose (sp gr 1.10) as described in the ICR method. The resulting pellet material was combined and transferred to a Leighton tube for further isolation of *C. parvum* oocysts using IMS.

4.3 Results and Discussion

In trials using 5 g of St. Albans digester sludge, with approximately 10% solids, direct IMS yielded oocyst recoveries of 21.3–52.1% with a spike dose of approximately 100 organisms (Table 4-1, trials 1, 2, and 3a). The mean oocyst recoveries for these trials were 43.6 ± 10.1%. Unlike raw sewage, the use of 0.75 g kaolin in sewage sludge did not improve IMS performance (Table 4-1, trial 3b). Increasing the sample size to 10 g or 15 g and analyzing as two subsamples using IMS decreased oocyst recoveries further (Table 4-2), suggesting that increasing particle concentrations probably impacted IMS performance.

Table 4-1. Direct Recovery of *C. parvum* Oocysts from St. Albans Sludge (5 g) Using IMS.

Trial (*n*)	Spike Dose(Mean ± SD)	Percent Recovery(Mean ± SD)
1 (5)	93.9 ± 8.0	41.1 ± 9.2
2 (5)	94.5 ± 10.1	48.3 ± 11.4
3a (5)	104.0 ± 12.3	43.1 ± 5.6
3b (5)	104.0 ± 12.3	30.8 ± 8.3

Trial 1: *C. parvum* oocysts spiked directly into Leighton tubes. In remaining trials oocysts were spiked into samples in individual glass beakers. In trial 3b, 0.75 g kaolin was added to each replicate.

Table 4-2. Recovery of *C. parvum* Oocysts in St. Albans Sludge Using a Modified IMS Procedure.

Weight of Sludge Analyzed, g (*n*)	Percent Recovery (Mean ± SD)
10 (6)	21.1 ± 5.4
15 (6)	7.6 ± 4.3

Mean spike dose: 88.8 ± 8.5 *C. parvum* oocysts.

To reduce particle concentration in samples, coarse filtration of the sludge followed by IMS was used; however, minimal oocyst numbers were detected in the supernatant (<1.0%). Also, where the initial sample weights were increased from 5 g to 25 g or 50 g, examination of the pellet material after coarse filtration indicated oocyst recoveries declining from 10.7% to 8.3%, and 1.5%, respectively (data not shown), providing further evidence for the effect of increased particle concentrations on IMS.

Rather than coarse filtration to remove large particles, slow-speed centrifugation was investigated as an alternative for separating larger particles while leaving oocysts in the supernatant. Several different centrifugation speeds were examined and indicated that low-speed centrifugation (100 x *g*) provided good oocyst recoveries; however, better separation between the solid and liquid phase was observed with 400 x *g*. For example, supernatant subsamples from 10 g sludge suspensions yielded oocyst recoveries of 29.5 ± 14.9 % at 100 x *g* centrifugation speed (data not shown) and oocyst recoveries declined to 8.5 ± 5.5% when centrifugation speed was increased to 400 x *g*. Similarly, in DI water better oocyst recoveries were seen in supernatant taken after centrifugation at 100 x *g* (33.7%) than at 400 x *g* (7.5%). These results suggested that even with slow-speed centrifugation (i.e., 100 x *g*; 2 min), a large proportion of the spike dose of oocysts settled into the pellet and as a result this approach for reducing particles in sludge samples was not a viable approach.

Using lime-stabilized ALCOSAN samples with considerably higher solids content (25–27%), it became difficult to get good mixing in the Leighton tubes. As a result, the Dynal mixing procedure was modified to use a flat plate magnet using a tilting orbital platform shaker. Despite efforts to improve mixing, the oocyst recoveries in both dewatered and lime-stabilized sludge were less than 2% (Table 4-3). In contrast, where St. Albans sludge was used with this mixing protocol, mean oocyst recoveries were 42.5% indicating that the mixing procedure was effective and the poor recoveries in the dewatered and lime-stabilized sludge were probably associated with the solids content of sample matrix.

Table 4-3. Oocyst Recoveries in Sludge (5 g) Using IMS and a Magnetic Panning Device.

Sludge (*n*)	*C. parvum* Oocyst Spike Dose (Mean ± SD)	Percent Recovery (Mean ± SD)
St. Albans (5)	91.3 ± 9.4	42.5 ± 3.0
ALCOSAN dewatered (2)	85.4 ± 9.1	1.2 ± 1.7
ALCOSAN lime stabilized (5)	91.3 ± 9.4	0.0 ± 0.0

To reduce further the solids content in samples, their partial purification before IMS was investigated by using flotation procedures (sp gr 1.10). A reduction in oocyst recoveries was noted in St. Albans sludge compared to direct IMS on 5 g samples (Table 4-4); however, where 2 g of dewatered sludge samples were used, oocyst recoveries improved to 35.7%. In contrast, the lime-stabilized sludge did not demonstrate improvements in recoveries by this procedure. In these samples, a gelatinous floc was noted after flotation, which may have interfered with oocyst isolation during IMS. These observations suggest that in addition to the concentration of particles, chemical composition of the sludge may also be a factor affecting IMS performance. The effects of dissolved iron on oocyst recoveries have been documented (Yakub & Stadterman-Knauer, 2000) and it was reported that chelators such as ethylenediamine tetracetic acid (EDTA) could be used to reverse specific interference. We have also noted that kaolin can play a similar role for reversal of unidentified inhibitors in sewage influent samples. Although kaolin docs not appear to offer any beneficial effects for sludge samples, the use of EDTA or other chelators was not investigated.

Table 4-4. Oocyst Recovery in Sludge Using Density Gradient Flotation Followed by IMS.

Sludge (*n*)	Weight Analyzed (g)	*C. parvum* Oocyst Spike Dose (Mean ± SD)	Percent Recovery (Mean ± SD)
DI (2)	-	88.8 ± 9.3	64.8 ± 4.0
St. Albans (5)	5	88.8 ± 9.3	25.9 ± 6.1
ALCOSAN dewatered (5)	2	81.7 ± 8.5	35.7 ± 10.0
ALCOSAN lime-stabilized (5)	2	81.7 ± 8.5	0.0 ± 0.0[1]

[1]A gelatinous floc formed during the flotation procedure.

During these investigations, the White House draft method (U.S. EPA, 1992) for *Ascaris* ova recovery from sludge was modified to recover *C. parvum* oocysts from the sludge matrix. When the supernatant was captured on 0.45-µm pore membrane filters and organisms recovered from the membranes, recoveries were 0.7–5.2% with a mean oocyst recovery of 2.4 ± 1.5%. Oocyst recoveries were 0.4–1.8 % with a mean oocyst recovery of 1.0 ± 0.5 % when the supernatant was passed through a glass fiber filter and the filtrate was concentrated and subjected to IMS to recover the oocysts (data not shown). Concentrating the supernatant on PCTE membranes and stomaching filters to recover oocysts yielded highest oocyst recoveries (mean 4.4 ± 2.7%), with recoveries of 1.4–10.1%. Although the objective of this procedure was to determine whether larger amounts of sludge samples could be analyzed for the presence of target organisms, the final stage of analysis could only be performed on one-tenth of the original volume (or 50 mL). This method was not only cumbersome, but used various manipulations, resulting in numerous stages in the procedure for potential loss of oocysts.

Iacovski et al. (2004) reported detecting 1 to 4 indigenous oocysts in sewage samples (1.0-, 2.0-, and 3.0-g sample sizes) and *Color*Seed recoveries of 40.0 ± 11.2% in 27 samples tested. The authors allowed the sludge suspension to settled and analyzed the supernatant for oocysts. They found that 1-g samples offered the highest oocyst recovery rates (46.5 ± 8.1%) and maximum oocyst recoveries were obtained with a 20-min sedimentation (43.0 ± 8.3%). Although this study used different modifications to either increase sample size for analysis or to improve overall recoveries, they did not improve. Other techniques, such as slow-speed centrifugation and coarse filtration, also failed to improve oocyst recoveries from biosolids samples. The only procedure that showed acceptable oocyst recoveries relied on using small sample weights (<5 g). Addition of chemical agents, such as EDTA or NaHMP, may improve oocyst recoveries in biosolids although these trials were not conducted because of time constraints. In addition, trials using *Color*Seed in biosolids were not conducted however, incorporating it into the assay as an internal control for method performance is possible.

CHAPTER 5.0

CRYPTOSPORIDIUM OCCURRENCE IN 10 WWTPS IN THE UNITED STATES

5.1 Introduction

Occurrence surveys to determine oocyst concentrations in wastewater have used a variety of methods for concentration and isolation of the oocysts from the contaminating debris. These methods have included modifications of the ICR or SCA methods (Enriquez et al., 1995; Medema et al., 1998; Jolis et al., 1999; Carraro et al., 2000), calcium carbonate flocculation followed by cold sucrose flotation (Payment et al., 2001), and coarse filtration of small volumes (50–2000 mL) through muslin followed by centrifugation and purification using various density gradients and purification solutions (Bukhari et al., 1997; Robertson et al., 2000). Jolis et al. (1999) detected no oocysts in reclaimed water using the ICR method. Although recovery efficiencies of the ICR method were not assessed in this Jolis et al. (1999) study, Clancy et al. (1994) reported poor recoveries of both *Cryptosporidium* oocysts and *Giardia* cysts, generally less than 10%, when using this method. Carraro et al. (2000) reported detecting *Cryptosporidium* oocysts at concentrations of 4.5 ± 0.8/L in raw influents ($n = 3$) and 0.21 ± 0.06% in effluents ($n = 11$) when using the ICR method, but again method recovery was not assessed. Enriquez et al. (1995) assessed method recoveries using a modification of the ICR method in DI and reclaimed water. These modifications included increasing the sample collection flow rate to 15–19 L/min (from 4 L/min), splitting the concentrated pellet in half, and preserving one half with potassium dichromate for detection of *Cryptosporidium* oocysts. This subsample was subjected to flotation with sucrose (sp gr 1.24) and the interface harvested and applied to cellulose-acetate membrane filters. The mean recovery rate for *Cryptosporidium* in tap water was 9.4 ± 1.4% and 8.7 ± 0.5% in reclaimed water.

Skerrett and Holland (2000) assessed recoveries using the calcium carbonate flocculation technique with and without cold sucrose flotation in DI water (1 L) and lake water (10 L). Reported recoveries in 1 L DI water of 32–49% without flotation and dropped to 16–29% with isolation of the oocysts using cold sucrose flotation. Oocyst recoveries in 1 L lake water were better than DI water, from 48% to 86% without flotation. Recoveries in 1 L lake water dropped slightly, to 32–56%, when sample concentrates were subjected to cold sucrose flotation. Similar recoveries in 10 L lake water samples processed with sucrose flotation were noted with spike doses of 100–1000 oocysts. In this study, river samples rather than wastewater were analyzed. However, one site was located downstream from a WWTP's discharge pipe and another site was located downstream from a cattle and sheep grazing area. *C. parvum* oocysts were detected at concentrations of <1/L–4/L at the former site and 1–1.5/L at the latter.

Bukhari et al. (1997) reported *C. parvum* oocyst concentrations in wastewater influents and effluents collected at seven treatment plants. Oocysts were detected in raw influents from

10–170/L when samples were subjected to coarse filtration, centrifugation, sucrose flotation, and further concentration using water and diethyl ether. Oocysts were detected in 27.2% of the raw influents analyzed ($n = 81$). Sewage effluents were not subjected to flotation or other purification techniques and were examined directly after concentration by centrifugation. Oocyst concentrations in effluents were reported to be 10–60/L with 25.5% of samples analyzed positive for *C. parvum* oocysts ($n = 94$).

Robertson et al. (2000) conducted a study of *Cryptosporidium* oocyst and *Giardia* cyst recovery methods using various purification techniques in wastewater in Scotland over 3 years. The mean recovery rate for *C. parvum* oocysts in sewage influents at low doses (~120 oocysts/L) were reported highest when small volumes were subjected to centrifugation followed by direct staining of the pellet material at 81 ± 21% ($n = 8$). The mean recovery rates fell to less than 35% when the pellet was subjected to additional manipulations to isolate the oocysts from the contaminating debris. Similar reductions in mean oocyst recoveries were noted when these isolation/purification techniques were used to recover oocysts from sewage effluents. The mean oocyst recovery rate fell from 89 ± 10% ($n = 8$) using direct examination of the concentrate to less than 45% using the three isolation techniques evaluated. After recoveries were assessed, the researchers evaluated the oocyst concentrations at six WWTPs in Scotland using each concentration/isolation technique. Reported levels of indigenous oocysts at five of six plants were <20–400/L in sewage influents when samples were concentrated and examined without further processing. These levels dropped to less than 20–194/L when further manipulations were conducted. The concentration of oocysts at the sixth plant was 10–6000/L when subjected to various isolation/purification technique postconcentration; however, oocyst levels in the centrifuged pellet were not determined at this plant suggesting that the level of solids present in the pellet was too high to examine directly. In effluent samples Robertson et al. (2000) reported oocysts levels of <20–1000/L when samples concentrates were examined directly. These levels dropped to <20–205/L when sample concentrates were subjected to purification/isolation techniques.

In the occurrence survey in this study, 10 WWTPs across the United States sent monthly grab or composite samples to the laboratory for analysis. Raw influents and primary effluents were collected as 1 L grab or composites samples. Secondary and tertiary effluent samples were filtered in the field, with *Color*Seed added to the filter capsule before sample collection. *Color*Seed was incorporated as an internal control to indicate method performance in each sample. It was not always possible for each utility to submit samples every month and not all utilities began participating in the survey at the same time. Participating utilities were located in Alabama, California, Colorado, North Carolina, Pennsylvania, and Vermont. Indigenous oocyst concentrations of <2–86/L in raw influents and primary effluents and <0.008–40.8/L in secondary and tertiary effluents. *Color*Seed recoveries were 0–83.8% across all wastewater matrices tested in the survey.

5.2 Materials and Methods

5.2.1 A Description of WWTPs

5.2.1.1 Alabama Plant #1

This plant operates at 16–18 million gallons per day (mgd) with a peak design flow of 35 mgd. The wastewater influent is comprised of domestic waste (70%) and industrial waste (30%

bakery and chicken/cattle slaughter houses). Influent wastewater is pretreated using screening and grit removal and enters the primary clarifiers. The primary weir effluent proceeds to the secondary clarifiers where more settling occurs for particle removal (aeration, 6 h detention). The final effluent remains in a chlorine contact basin for 30 min before being discharged to receiving waters. The chlorine residual in the final effluent is approximately 0.1 part per million (ppm) with a maximum residual limit of 0.37 ppm.

5.2.1.2 Alabama Plant #2

This plant operates at 10–12 mgd with a peak flow design of 21 mgd. The wastewater influent is comprised of 60% domestic and 40% industrial (e.g., surgical implants manufacturing, sheet metal, food industry). Preliminary treatment of the influent includes screen and grit removal as well as pre-aeration. The pretreated wastewater enters the primary clarifiers and the weir effluent is passed through trickling filters. The trickling filter effluent enters the secondary clarifiers. The effluent from the secondary clarifiers remains in a chlorine contact basin for 30 min before being discharged into receiving waters.

5.2.1.3 Alabama Plant #3

This plant operates at 2 mgd with a peak flow design of 3 mgd. The wastewater influent is comprised of 70% domestic and 30% industrial (food industry). Preliminary treatment of the influent includes screen and grit removal as well as pre-aeration. The pretreated wastewater enters the primary clarifiers and the weir effluent is passed through biofilters, intermediate clarifiers, and then trickling filters. The trickling filter effluent enters the secondary clarifiers. The effluent from the secondary clarifiers remains in a chlorine contact basin for 30 min before being discharged.

None of the Alabama plants used chemicals, such as polymers, alum, or caustic for enhanced coagulation, phosphorus removal or pH adjustment. All Alabama plants are located in the same city.

5.2.1.4 Vermont Plant

This advanced aeration plant has an average flow of 0.6 mgd with a peak flow design of 1.25 mgd. Wastewater influent is comprised of domestic waste (90%), and effluent from a microbrewery. The raw influent travels directly to the aeration tanks after grit removal. The aeration tank outfall is treated with alum, which is applied at a dose of 5 mg/L for phosphorus removal and then flows on to the secondary clarifiers. The clarifier effluent (secondary effluent) is passed through a cloth filter (10–20 microns) and passes by two banks of UV lights (to reduce the levels of fecal coliform bacteria) before exiting the plant.

5.2.1.5 Colorado Plant

This plant currently operates at an average daily flow of 12 mgd with a maximum flow rate of 17 mgd. The average design flow is 23 mgd with a peak hour flow capacity of 35.6 mgd. The wastewater influent is 96% domestic waste with brewery waste comprising the balance of the influent. The treatment process is conventional activated sludge plant. The secondary clarifier effluent is chlorinated and dechlorinated before discharge from the plant.

5.2.1.6 California Plant

This plant receives 95% domestic and 5% light industrial wastewater with an average daily flow of 17 mgd and a peak daily flow of 30 mgd. This fully combined sewer system may reach peak flows of 65 mgd during the wet weather season. During this time 43 mgd receives secondary treatment and the remainder only primary. The primary effluent is blended with the discharge effluent providing the flow is ≤65 mgd. If the flow exceeds 65 mgd, decanted effluent from the outfall is discharged with the plant effluent. Ferric chloride is added to the secondary clarifiers to aid in the coagulation/flocculation process. In the summer hypochlorite is added to the influent stream to aid in odor control.

5.2.1.7 North Carolina Plant

This plant is a reclamation facility with an average of 8% of the daily flow being diverted for irrigation use. Primary treatment involves the use of a two-stage phase bar screen and grit removal. A phased isolation ditch provides biological nutrient reduction by alternating the wastewater stream between aerobic and slightly anaerobic conditions to convert ammonia to nitrogen gas and lower the levels of biological oxygen demand and phosphorus without addition of chemicals. The wastewater stream flows to the clarifiers for reduction of particles. Clarifier effluent is passed through sand filters for additional particle/biological removal, with the effluent treated with UV light (average dose, 40 mJ/cm^2). Additional disinfection is achieved by dosing with hypochlorite to maintain a residual of 2 ppm. Average daily flow through this plant is 7 mgd with a peak flow capacity of 12 mgd. Domestic wastewater comprises 93 % of the wastewater with the balance being pharmaceutical and food processor (cookies, crackers) waste.

5.2.1.8 Pennsylvania Plant #1

This plant is a conventional secondary treatment facility receiving primarily domestic waste. The average daily flow is 186 mgd with a maximum flow of 420 mgd. Phosphorus is added to the wastewater stream before aeration to provide nutrients for the microbial population. Sludge received from a drinking water plant contains high levels of ferric chloride that depletes the available phosphorus in the wastewater. The secondary effluent is chlorinated before pumping the discharge to a river.

5.2.1.9 Pennsylvania Plant #2

This plant is a conventional secondary treatment facility receiving primarily domestic waste. The average daily flow is 93 mgd with a design flow of 292 mgd. Phosphorus is added to the wastewater stream before aeration to provide nutrients for the microbial population. Sludge received from a drinking water plant contains high levels of ferric chloride that depletes the available phosphorus in the wastewater. The secondary effluent is chlorinated before discharging into a river.

5.2.1.10 Pennsylvania Plant #3

This plant receives primarily domestic waste with about 10% being industrial. This plant uses high purity oxygen in the primary clarifiers to maximize the oxidation rate of iron and manganese in the wastewater. The average daily flow is 193 mgd with a peak daily flow of 451mgd. The secondary effluent is chlorinated and then discharged to a river.

All Pennsylvania plants are located in the same city.

5.2.2 Protocols for Analysis of Wastewater Samples

Protocols provided as Appendices A and B in this report were followed for collection and analysis of all wastewater matrices tested; raw influent, primary effluent, secondary effluent and tertiary effluent. Briefly, sample instructions, data sampling sheets, sample bottles, HV capsule filters, *Color*Seed, and 0.05% Tween 80 and coolers with ice packs were sent to utility personnel. One utility did not have the equipment or resources to filter their secondary effluent samples and collected grab samples in cubitainers and shipped them on ice to CEC. Raw influent and primary effluent samples were collected as 1 L samples. Just before collection of the secondary and tertiary effluents, the contents of the *Color*Seed vial was transferred to the HV capsule filter in the field. The secondary effluent was drawn through the filter until a volume of approximately 10 L was collected. Tertiary effluent was filtered through the HV filter until the flow rate dropped to less than 0.5 L/min. Samples were transported to the laboratory by overnight courier or on blue ice by laboratory personnel (Vermont samples) and analyzed within 24 hours of receipt.

5.3 Results and Discussion

5.3.1 Wastewater Samples

5.3.1.1 Alabama Plants

Table 5-1 presents *Color*Seed recovery data for raw influent and primary effluent samples collected at Alabama Plant #1 from May 2002 through April 2003. Indigenous oocysts were detected in 54.5% (6/11) of raw influent samples and 72.7% (8/11) primary effluent samples analyzed. Concentration of indigenous oocysts were <2–24/L in raw influent and <2–8/L in primary effluent. *Color*Seed recoveries were 17.2–83.8% in raw influent with a mean *Color*Seed recovery of 44.2 ± 18.0% ($n = 11$). *Color*Seed recoveries in primary effluents were slightly higher than in raw influent samples at 48.9 ± 9.9% ($n = 11$). *Color*Seed recoveries were 34.3–65.7% in primary effluent samples.

Results for Alabama Plant #2 are presented in Table 5-2. Mean *Color*Seed recoveries were slightly lower than observed at Alabama Plant #1 (35.0 ± 18.2%) with recoveries of 12.1–59.6% ($n = 11$). *Color*Seed recoveries in primary effluent samples at Plant #2 were 8.1–78.8% ($n = 11$) with a mean *Color*Seed recovery of 44.9 ± 19.6%. Indigenous oocysts were detected in 18.2% (2/11) of raw influent samples collected at concentrations of <2–2/L. Concentrations of indigenous oocysts in primary effluents were <2–82/L.

*Color*Seed recoveries in raw influent and primary effluent samples collected at Alabama Plant #3 are presented in Table 5-3. These results show the lowest mean *Color*Seed recoveries in both raw influent and primary effluent samples of the three plants in Alabama. *Color*Seed recoveries in raw influent samples were 0–46.5% with a mean result of 23.4 ± 15.9%. Indigenous oocysts were detected in three of 11 raw influent samples (27.3%) at concentrations of <2–2/L. In primary effluent samples, 18.2% (2/11) were positive for indigenous oocysts.

Results from the analyses of secondary effluent samples collected at all three Alabama plants are presented in Table 5-4. In each case, 10 L samples were collected on one Envirochek HV filter with the exception of the July sample, collected at Plant #3, when two filters were required to collect a total volume of 10 L. *Color*Seed recoveries were less than 10% on: seven of 11 occasions (at Plant #1), five out of 11 (at Plant #2), and one out of 11 (at Plant #3). Mean

Table 5-1. *Color*Seed Recoveries and Number of Indigenous Oocysts Detected in Raw Wastewater Influents and Primary Effluents (500 mL samples) from Alabama Plant #1 from May 2002 through April 2003.

Matrix	Month	Turbidity (ntu)	Packed Pellet Volume (mL)	*Color*Seed[1] % Recovery	Number of IO[2] Detected in 500 mL Sample Volume Examined
Raw Influent	May	138	0.75	19.4	0
	June	NR[3]	1.0	52.5	2
	July	117	0.5	42.4	0
	August	103	0.8	41.4	2
	September	152	1.0	17.2	0
	October	89.5	0.4	83.8	12
	November	116	0.4	50.5	1
	December	86.0	0.75	44.4	1
	January	155	0.5	33.0	0
	February		Samples not collected during the month of February		
	March	86.3	0.4	50.5	0
	April	52.9	0.5	51.5	4
	Mean *Color*Seed Recovery ± SD			**44.2 ± 18.0**	
Primary Effluent	May	64.9	0.2	62.2	1
	June	NR[3]	0.2	41.4	0
	July	41.3	0.2	48.5	4
	August	67.8	0.1	65.7	1
	September	57.8	0.2	34.3	4
	October	47.4	0.2	50.5	1
	November	49.2	0.2	40.4	0
	December	65.3	0.1	48.5	3
	January	68.2	0.1	43.0	0
	February		Samples not collected during the month of February		
	March	70.1	0.2	59.6	1
	April	41.4	0.2	43.4	1
	Mean *Color*Seed Recovery ± SD			**48.9 ± 9.9**	

1. *Color*Seed spike dose concentration: 99–100 oocysts.
2. IO, indigenous oocysts.
3. NR, not recorded.

*Color*Seed recovery at plant #1 was 18.6 ± 21.9% with recoveries of 2.0–63.0%. The mean *Color*Seed recovery at Plant #2 was 19.5 ± 18.9%, with recoveries of 0–47.5%. At Plant #3 mean *Color*Seed recovery was 27.3 ± 12.1% with recoveries of 9.1–47.5%. Indigenous oocysts were detected in more than 50% of the secondary effluent samples collected at each plant. At Plant #1, six of 11 (54.5%) samples contained indigenous oocysts. On two of these occasions the *Color*Seed recoveries were less than 5%. Indigenous oocysts were detected at Plant #2 in 63.6% of samples collected and in 72.7% of those collected at Plant #3.

Concentration of indigenous oocysts for all samples at Plant #1 were <1–1/L, <1–14.5/L at Plant #2, and <1–0.4/L at Plant #3. Mean indigenous oocyst concentration was highest at Plant #2 (22.6 ± 44.7) where on three occasions the oocyst concentrations exceeded 1/L and was as high as 14.5/L (June). In one instance, indigenous oocysts were detected in the sample although no *Color*Seed oocysts were recovered. These results suggest that matrix composition had inhibitory properties that prevented the recovery or isolation of the *Color*Seed oocysts; however, the concentration of indigenous oocysts may have been sufficiently high so that some were

Table 5-2. *Color*Seed Recoveries and Number of Indigenous Oocysts Detected Detected in Raw Wastewater Influents and Primary Effluents (500 mL samples) from Alabama Plant #2 from May 2002 through April 2003.

Matrix	Month	Turbidity (ntu)	Packed Pellet Volume (mL)	*Color*Seed[1] % Recovery	Number of IO[2] Detected in 500 mL Sample Volume Examined
Raw Influent	May	254	0.75	19.4	0
	June	NR[3]	0.8	59.6	0
	July	279	2.0	34.3	0
	August	700	4.0	12.1	0
	September	126	0.5	32.3	0
	October	119	0.5	73.7	1
	November	492	1.5	17.2	0
	December	194	0.5	32.3	1
	January	269	1.0	42.0	0
	February		Samples not collected during the month of February		
	March	202	1.0	34.3	0
	April	101	0.8	28.3	0
	Mean *Color*Seed Recovery ± SD			**35.0 ± 18.2**	
Primary Effluent	May	81.4	0.4	43.9	0
	June	NR[3]	0.4	42.4	41
	July	80.4	0.4	54.5	5
	August	120	0.4	78.8	15
	September	56.2	0.1	8.1	0
	October	67.7	0.3	56.6	0
	November	91.7	0.2	31.3	0
	December	96.8	0.4	44.4	1
	January	96.8	0.2	63.0	0
	February		Samples not collected during the month of February		
	March	79.0	0.2	49.5	0
	April	55.2	0.2	21.2	1
	Mean *Color*Seed Recovery ± SD			**44.9 ± 19.6**	

1. *Color*Seed spike dose concentration: 99–100 oocysts.
2. IO, indigenous oocysts.
3. NR, not recorded.

recovered in the IMS step. Plant #1 receives 30% of its wastewater from various food industries (cattle and poultry slaughterhouses, bakeries) and at Plant #2, 40% of the waste is from industries such as sheet metal and surgical implant manufacturing, as well as the food industry. The chemicals used to clean equipment in the food processing industries and metals that leach into the waste stream from the metal industries may produce chemical interferences that inhibit the performance of the IMS kit.

5.3.1.2 Vermont Plant

Wastewater samples collected at an advanced aeration plant with tertiary treatment located in Vermont are presented in Table 5-5. No indigenous oocysts were detected in any of the samples collected with the exception of one raw influent sample (February). *Color*Seed recoveries in the raw influent were 17.2–72.0% with a mean *Color*Seed recovery of 42.0 ± 14.0% ($n = 11$). Mean *Color*Seed recovery in the secondary effluent was 30.4 ± 13.7% with recoveries of 2.0–50.5%. For the sample collected in May, the *Color*Seed recovery was 2.0%; however, during sample collection some of the contents of the capsule were lost. This may account for the poor recovery result noted this month. The mean *Color*Seed recovery without

Table 5-3. *Color*Seed Recoveries and Number of Indigenous Oocysts Detected in Raw Wastewater Influents and Primary Effluents (500 mL samples) from Alabama Plant #3 from May 2002 through April 2003.

Matrix	Month	Turbidity (ntu)	Packed Pellet Volume (mL)	*Color*Seed[1] % Recovery	Number of IO[2] Detected in 500 mL Samples Examined
Raw Influent	May	208	1.0	23.5	0
	June	NR[3]	1.2	43.4	1
	July	563	4.0	0	0
	August	557	5.0	11.1	0
	September	404	2.0	9.1	0
	October	79.1	0.5	43.4	1
	November	405	1.5	17.2	0
	December	302	0.75	24.2	1
	January	264	0.5	9.0	0
	February		Samples not collected during the month of February		
	March	269	0.2	46.5	0
	April	207	1.0	30.3	0
	Mean *Color*Seed Recovery ± SD			**23.4 ± 15.9**	
Primary Effluent	May	23.5	0.2	48.0	1
	June	NR[3]	0.4	23.2	0
	July	12.0	0.2	37.4	0
	August	30.7	0.3	52.5	0
	September	15.8	0.1	5.1	0
	October	10.9	0.1	32.3	0
	November	21.6	0.1	51.5	1
	December	49.2	0.2	48.5	0
	January	23.5	0.2	25.0	0
	February		Samples not collected during the month of February		
	March	23.5	0.5	50.5	0
	April	18.6	0.2	33.3	0
	Mean *Color*Seed Recovery ± SD			**37.0 ± 15.1**	

1. *Color*Seed spike dose concentration: 99–100 oocysts.
2. IO, indigenous oocysts.
3. NR, not recorded.

this data point is 33.3 ± 10.5%. Samples of tertiary effluent were 14.8–131.3 L. The sampler allowed collection of the tertiary effluent until the filter plugged. A long period of subfreezing weather in January and February prevented adequate particle removal in the wastewater treatment process preventing the collection of higher volumes of tertiary wastewater. Mean *Color*Seed recovery in tertiary effluent was 51.8 ± 7.7% with oocyst recoveries of 35.4–62.2%.

5.3.1.3 Colorado Plant

Table 5-6 presents data from wastewater samples collected at the secondary treatment facility in Colorado. No indigenous oocysts were detected in the raw influent samples analyzed. In five of 9 raw influent samples, *Color*Seed recoveries were less than 10%. The mean *Color*Seed recovery was 11.8 ± 8.7% with recoveries of 3.0–25.3%. *Color*Seed recoveries in

Table 5-4. *Color*Seed Recoveries in 10-L Samples of Secondary Effluent from Three Alabama WWTPs from May 2002 through April 2003.

Plant #	Month	Turbidity (ntu)	Packed Pellet Volume (mL)	*Color*Seed[1] % Recovery	Number of IO[2] Detected in 10 L Sample Examined
1	May	4.63	0.5	2.0	0
	June	1.38	0.5	9.1	0
	July	1.37	0.4	9.1	0
	August	3.08	0.4	6.1	0
	September	1.65	0.6	22.2	3
	October	2.68	0.2	2.0	0
	November	3.10	0.2	3.0	2
	December	1.84	0.2	32.3	10
	January	1.05	0.2	53.5	5
	March	1.41	0.2	63.0	7
	April	1.87	0.5	2.0	3
Mean *Color*Seed Recovery at Plant #1				**18.6 ± 21.9%**	
2	May	17.4	1.5	0	0
	June	13.3	1.0	34.3	145
	July	3.80	2.0	3.0	0
	August	12.9	0.4	26.3	3
	September	9.82	0.9	26.3	6
	October	10.6	1.0	32.3	0
	November	12.2	0.5	47.5	6
	December	19.3	0.3	43.4	63
	January	11.0	0.5	0	24
	March	12.3	1.0	0	0
	April	9.63	0.8	1.0	2
Mean *Color*Seed Recovery at Plant #2				**19.5 ± 18.9%**	
3	May	16.1	1.5	32.7	15
	June	7.27	1.0	33.3	3
	July	18.9	0.5	13.1	0
	August	2.90	0.2	47.5	10
	September	5.00	0.6	24.2	2
	October	6.34	1.0	42.4	12
	November	9.26	0.8	23.2	2
	December	13.2	2.0	33.3	36
	January	10.3	1.0	9.1	0
	March	6.43	1.0	27.0	4
	April	8.21	0.8	14.1	0
Mean *Color*Seed Recovery at Plant #3				**27.3 ± 12.1%**	

1. *Color*Seed spike dose concentrations: 99 or 100.
2. IO, indigenous oocysts.

primary effluent were 1.0–51.0% with a mean recovery of 23.0 ± 13.8%. Indigenous oocysts were detected in 33.3% (3 of 9) primary effluent samples analyzed. Concentrations of indigenous oocysts detected were <2–24/L. In secondary effluent indigenous oocysts were detected in all samples analyzed with concentrations of 0.1–0.7/L. Indigenous oocyst concentrations were not determined for each matrix for samples collected in July. Live oocysts were added to the samples for cell culture trials (results presented in Chapter 6.0).

Table 5-5. *Color*Seed Recoveries and Number of Indigenous Oocysts Detected in Wastewater Samples Collected at a Vermont Advanced Aeration Plant with Tertiary Treatment from May 2002 through April 2003.

Matrix	Month	Sample Volume Analyzed (L)	Turbidity (ntu)	Packed Pellet Volume (mL)	*Color*Seed[1] % Recovery	Number of IO[2] Detected
Raw Influent	May	0.5	163	1.0	40.8	0
	June	0.5	116	1.0	40.4	0
	July	0.5	72.9	0.7	48.5	0
	August	0.5	163	1.5	44.4	0
	September	0.5	241	0.5	38.4	0
	October	0.5	166	2.0	17.2	0
	November	0.5	86.6	0.5	51.5	0
	December	0.5	296	0.5	43.4	0
	January	0.5	114	1.0	25.0	0
	February	0.5	72.3	0.5	72.0	11
	April	0.5	124	0.5	40.4	0
	Mean *Color*Seed Recovery ± SD				**42.0 ± 14.0**	
Secondary Effluent	May	12.0	5.85	1.0	2.0[3]	0
	June	9.8	2.55	1.0	20.2	0
	July	10.0	2.71	0.4	32.3	0
	August	11.4	1.60	0.75	21.2	0
	September	11.4	3.23	0.5	50.5	0
	October	10.6	2.78	1.0	38.4	0
	November	10.2	4.00	0.3	49.5	0
	December	9.8	1.74	0.4	33.3	0
	January	10.0	4.33	0.5	33.0	0
	February	10.0	10.9	0.5	29.0	0
	April	10.0	6.97	2.0	25.3	0
	Mean *Color*Seed Recovery ± SD				**30.4 ± 13.7**	
Tertiary Effluent	May	60.6	1.46	1.0	59.2	0
	June	46.9	2.14	0.4	47.5	0
	July	100	2.16	0.5	35.4	0
	August	85.7	0.84	0.2	52.5	0
	September	60.6	1.99	0.1	62.6	0
	October	99.9	0.77	0.2	50.5	0
	November	43.1	2.43	0.2	51.5	0
	December	83.5	1.06	0.1	57.6	0
	January	14.8	1.99	0.3	56.0	0
	February	18.9	1.37	0.2	54.0	0
	April	131.3	2.62	1.0	43.4	0
	Mean *Color*Seed Recovery ± SD				**51.8 ± 7.7**	

1. *Color*Seed spike dose concentrations: 99 or 100.
2. IO, indigenous oocysts.
3. Low recovery caused by losing capsule contents during sampling.

Table 5-6. *Color*Seed Recoveries and Number of Indigenous Oocysts Detected in Wastewater Samples Collected at a Colorado Secondary Treatment Plant from October 2002 through July 2003.

Matrix	Month	Sample Volume Analyzed (L)	Turbidity (ntu)	Packed Pellet Volume (mL)	*Color*Seed[1] % Recovery	Number of IO[2] Detected
Raw Influent	October	0.5	248	2.0	17.2	0
	November	0.5	166	1.0	6.1	0
	December	0.5	260	2.0	5.1	0
	January	0.5	261	2.0	4.0	0
	February	0.5	254	0.5	25.0	0
	March	0.5	192	0.5	3.0	0
	April	0.5	234	0.5	9.1	0
	May	0.5	162	0.5	25.3	0
	July	0.5	89.2	0.5	11.2	ND[3]
Mean *Color*Seed Recovery ± SD					**11.8 ± 8.7**	
Primary Effluent	October	0.5	156	0.2	31.3	0
	November	0.5	89.9	0.2	27.3	0
	December	0.5	95.8	0.2	16.2	1
	January	0.5	102	0.2	22.0	0
	February	0.5	97.0	0.2	51.0	0
	March	0.5	139	0.5	25.3	1
	April	0.5	120	0.5	13.1	12
	May	0.5	70.4	0.2	1.0	0
	July	0.5	42.7	0.5	20.2	ND[3]
Mean *Color*Seed Recovery ± SD					**23.0 ± 13.8**	
Secondary Effluent	October	13.2	3.50	1.0	51.5	3
	November	10	4.10	0.8	11.1	5
	December	9.5	5.90	1.0	7.1	2
	January	10.2	4.85	1.0	36.4	2
	February	10	2.10	0.75	52.0	1
	March	9.8	3.10	1.0	47.5	1
	April	9.8	4.20	1.0	6.1	2
	May	9.8	4.21	0.5	30.3	7
	July	10	2.40	0.5	32.3	ND[3]
Mean *Color*Seed Recovery ± SD					**30.5 ± 18.5**	

1. *Color*Seed spike dose concentrations: 99 or 100.
2. IO, indigenous oocysts.
3. ND, not determined.

5.3.1.4 California Plant

*Color*Seed recovery data for a secondary treatment plant in California are presented in Table 5-7. Sufficient sample volume of each matrix was collected in May to examine variability of the matrix. *Color*Seed recoveries in the raw influent were less than 15% with the exception of one sample analyzed in July where the *Color*Seed recovery was 27.8%. Triplicate subsamples analyzed in May showed very consistent *Color*Seed recoveries and the number of indigenous oocysts detected at 7.1 ± 0% ($n = 3$) and <2/L, respectively. Indigenous oocysts were detected in 5 of 11 raw influent samples analyzed and were not determined in the July sample. The *Color*Seed recoveries in primary effluents analyzed were much improved over the raw influent results with recoveries were 0–68.7% and a mean of 37.3 ± 20.2%. However, a high level of variability was noted in the triplicate subsamples analyzed in May. *Color*Seed recoveries were 0–68.7%, while concentrations of indigenous oocysts detected were <2–24/L in these replicate samples. Indigenous oocysts were detected in all other primary effluent samples analyzed from this plant. The concentrations detected were 2–86/L. The highest level of indigenous oocysts detected in the primary effluents was from a sample collected in December, which is the same month that the highest levels were noted in the secondary effluent at 39/L. Indigenous oocyst concentrations in the secondary effluents were 0.1–39/L. The mean *Color*Seed recovery for the

Table 5-7. *Color*Seed Recoveries and Number of Indigenous Oocysts Detected in Wastewater Samples Collected at a California Secondary Treatment Plant from October 2002 through July 2003.

Matrix	Month	Sample Volume Analyzed (L)	Turbidity (ntu)	Packed Pellet Volume (mL)	*Color*Seed[1] % Recovery	Number of IO[2] Detected
Raw Influent	October	0.5	240	5.0	6.1	12
	November	0.5	142	2.0	5.1	1
	December	0.5	139	1.0	13.1	5
	January	0.5	211	1.0	0	0
	February	0.5	269	4.0	0	0
	March	0.5	222	5.0	12.0	0
	April	0.5	121	2.0	14.1	3
	May	0.5	113	1.0	7.1	0
	May	0.5	113	1.0	7.1	0
	May	0.5	113	1.0	7.1	0
	June	0.5	191	0.6	8.1	1
	July	0.5	35.3	0.5	27.8	ND[3]
	Mean *Color*Seed Recovery ± SD				**9.0 ± 7.4**	
Primary Effluent	October	0.5	61.6	0.2	22.2	22
	November	0.5	36.3	0.2	39.4	6
	December	0.5	80.9	0.1	39.4	43
	January	0.5	45.3	0.2	62.0	7
	February	0.5	58.1	0.2	32.0	1
	March	0.5	58.1	0.2	65.7	26
	April	0.5	58.1	0.2	34.3	4
	May	0.5	56.2	0.2	0	1
	May	0.5	56.2	0.2	36.4	12
	May	0.5	56.2	0.2	68.7	0
	June	0.5	48.3	0.2	28.3	6
	July	0.5	73.1	0.5	18.7	ND
	Mean *Color*Seed Recovery ± SD				**37.2 ± 20.2**	
Secondary Effluent	October	11.4	2.38	1.0	58.6	0
	November	11.2	2.26	2.0	19.2	0
	December	10.5	4.77	0.5	62.6	409
	January	10.0	6.67	0.5	1.0	4
	February	8.0	11.5	4.0	9.0	17
	March	10.5	1.47	1.5	5.0	60
	April	11.0	6.38	1.5	23.2	190
	May	10.5	6.58	0.6	16.2	20
	May	9.5	6.58	0.5	21.2	18
	May	10.5	6.58	0.5	28.3	18
	June	11.0	5.98	0.5	32.3	48
	July	11.5	11.5	0.8	21.2	ND
	Mean *Color*Seed Recovery ± SD				**24.8 ± 19.0**	

1. *Color*Seed spike dose concentrations: 99 or 100.
2. IO, indigenous oocysts.
3. ND, not determined.

secondary effluent samples analyzed was 24.8 ± 19.0% (n =12) and were 1.0–62.6%. Triplicate subsample analyses were more consistent than in primary effluent samples with *Color*Seed recoveries of 16.2–28.3% and indigenous oocyst concentrations of 1.7–1.9/L. These concentrations of indigenous oocysts in secondary effluent represent the highest detected in all samples analyzed for this project.

5.3.1.5 North Carolina Plant

*Color*Seed recoveries in raw influent samples collected at a tertiary treatment plant in North Carolina are presented in Table 5-8. These recoveries were 3.0–29.2% with a mean recovery of 16.7 ± 11.5%. Indigenous oocysts were only detected in one raw influent sample (March) at a concentration of 2/L. Mean *Color*Seed recovery in primary effluent was 11.5 ± 5.5% and no indigenous oocysts were detected in these samples. *Color*Seed recoveries in secondary effluent samples were 25.3–70.0% for a mean recovery result of 43.1 ± 17.6%. Indigenous oocysts were detected at concentration of <0.1–0.6/L. Mean *Color*Seed recovery in tertiary effluent was 44.0 ± 21.1%. This result is not different from the mean recovery result for the secondary effluent. Indigenous oocysts were detected in three of four tertiary samples. Interestingly, a portion of the effluent from this plant is used as irrigation water. Indigenous oocysts were not determined in the July samples as these samples were analyzed using cell culture techniques.

Table 5-8. *Color*Seed Recoveries and Number of Indigenous Oocysts Detected in Wastewater Samples Collected at a North Carolina Tertiary Treatment Plant from March 2003 through July 2003.

Matrix	Month	Sample Volume Analyzed (L)	Turbidity (ntu)	Packed Pellet Volume (mL)	*Color*Seed[1] % Recovery	Number of IO[2] Detected
Raw Influent	March	0.5	529	3.0	23.2	1
	April	0.5	287	1.0	22.2	0
	May	0.5	157	1.0	3.0	0
	June	0.5	183	1.2	6.1	0
	July	0.5	149	1.0	29.2	ND[4]
		Mean *Color*Seed Recovery ± SD			**16.7 ± 11.5**	
Primary Effluent	March	0.5	327	2.0	9.1	0
	April	0.5	240	1.0	15.2	0
	May	0.5	147	0.8	8.1	0
	June	0.5	142	0.5	6.1	0
	July	0.5	130	1.0	19.2	ND
		Mean *Color*Seed Recovery ± SD			**11.5 ± 5.5**	
Secondary Effluent	March	11.4	NR[3]	0.8	25.3	0
	April	11.4	0.8	1.0	29.3	2
	May	22.7	1.4	0.3	46.5	1
	June	22.7	1.7	0.1	44.4	14
	July	22.7	1.0	0.5	70.0	ND
		Mean *Color*Seed Recovery ± SD			**43.1 ± 17.6**	
Tertiary Effluent	March	22.7	0.75	0.5	63.6	2
	April	26.5	0.77	0.4	19.2	6
	May	22.7	0.8	<0.1	65.7	0
	June	22.7	1.4	<0.1	45.5	3
	July	22.7	1.7	0.1	26.2	ND
		Mean *Color*Seed Recovery ± SD			**44.0 ± 21.1**	

1. *Color*Seed spike dose concentrations: 99 or 100.
2. IO, indigenous oocysts.
3. ND, not determined.
4. NR, not recorded.

Table 5-9. *Color*Seed Recoveries and Number of Indigenous Oocysts Detected in Raw Wastewater Influents and Primary Effluents (500-mL samples) from Pennsylvania Plant #1 from November 2002 through July 2003.

Matrix	Month	Turbidity (ntu)	Packed Pellet Volume (mL)	*Color*Seed[1] % Recovery	Number of IO[2] Detected in 500 mL Sample Volume Examined
Raw Influent	November	136	0.5	11.1	0
	December	124	0.75	10.1	0
	January	97.0	0.5	40.0	3
	March		Samples not collected from this site in March		
	April	61.4	0.5	19.2	1
	May	152	1.0	13.1	0
	June	82.9	0.4	22.2	0
	July	130	0.5	13.2	ND[3]
	Mean *Color*Seed Recovery ± SD			**18.4 ± 10.5**	
Primary Effluent	November	74.3	0.3	25.3	2
	December	59.3	0.5	21.2	2
	January	50.1	0.3	45.0	1
	March		Samples not collected from this site in March		
	April	37.9	0.4	25.3	0
	May	91.3	0.3	33.3	4
	June	33.4	0.3	30.3	2
	July	55.3	0.2	16.2	ND[3]
	Mean *Color*Seed Recovery ± SD			**28.1 ± 9.3**	

1. *Color*Seed spike dose concentration 99–100 oocysts.
2. IO, indigenous oocysts.
3. ND, not determined.

5.3.1.6 Pennsylvania Plants

Wastewaters from three plants in Pennsylvania were received from November 2002 through July 2003, although samples were not received from Plant #1 in March. All samples were subsamples from 24 h composites samples. Recovery data from these plants are presented in Tables 5-9 through 5-12. Triplicate subsample analyses were performed on the raw influent and primary effluent samples collected at Plant # 3 in May. Duplicate analyses of the secondary effluent were performed on the May samples. *Color*Seed recoveries in raw influent were lowest at Plant #1 with recoveries of 10.2-40.0% and a mean recovery of 18.4 ± 10.5% (n = 7). The mean *Color*Seed recovery in raw influents at Plant #2 was 28.7 ± 11.0% (n = 8). *Color*Seed recoveries were 14.1–48.5%. At Plant #3, *Color*Seed recoveries were 19.2–47.5%, yielding the highest mean *Color*Seed recovery of the three plants at 29.5 ± 8.1% (n = 10). The analysis of triplicate subsamples of raw influent in May yielded recoveries of 20.2–31.3% with a mean recovery of 26.3 ± 5.6%. Indigenous oocysts were detected at all plants with the highest concentrations noted at Plant #3. Indigenous oocyst concentrations were detected in seven of nine raw influent samples at Plant #3 with concentrations of <2–8/L. In contrast, indigenous oocysts were only detected in one of seven raw influent samples at Plant #2 and two of six at Plant #1.

At Pennsylvania Plant #1 the mean *Color*Seed recovery in primary effluents was 28.1 ± 9.3% (n = 7) with indigenous oocysts being detected in five of six samples. Concentrations of indigenous oocysts detected were <2–8/L in the primary effluent at Plant #1. At Plant #2 the concentration of indigenous oocysts was lowest of the Pennsylvania plants, with only one sample positive for indigenous oocysts. The November sample was lost in a laboratory accident so no

Table 5-10. *Color*Seed Recoveries and Number of Indigenous Oocysts Detected in Raw Wastewater Influents and Primary Effluents (500 mL samples) from Pennsylvania Plant #2 from November 2002 through July 2003.

Matrix	Month	Turbidity (ntu)	Packed Pellet Volume (mL)	*Color*Seed[1] % Recovery	Number of IO[2] Detected in 500 mL Sample Volume Examined
Raw Influent	November	90.1	0.5	30.3	0
	December	45.4	0.75	25.3	0
	January	59.2	0.5	41.0	2
	March	60.0	0.5	14.1	0
	April	57.5	0.5	48.5	0
	May	93.0	1.0	22.2	0
	June	101	0.5	25.3	0
	July	87.3	0.8	23.2	ND[3]
Mean *Color*Seed Recovery ± SD				**28.7 ± 11.0**	
Primary Effluent	November	35.6	0.2	LA[4]	LA[4]
	December	4.08	0.1	18.2	1
	January	21.3	0.4	39.0	0
	March	31.9	0.2	4.0	0
	April	30.9	0.2	28.3	0
	May	18.5	0.2	30.3	0
	June	178	<0.1	5.1	0
	July	21.4	0.2	26.2	ND[3]
Mean *Color*Seed Recovery ± SD				**21.6 ± 13.2**	

1. *Color*Seed spike dose concentration: 99– 100 oocysts.
2. IO, indigenous oocysts.
3. ND, not determined.
4. LA, laboratory accident.

data for this month are available. The mean *Color*Seed recovery at Plant #2 was 21.6 ± 13.2% and recoveries were 4.0–39.0%. At Plant #3 the mean *Color*Seed recovery was 29.1 ± 9.1% (n = 10). Recoveries were 15.2–44.4%. Indigenous oocysts were detected in half of the samples with concentrations of <2–8/L. In May triplicate analyses on the primary effluent yielded *Color*Seed recoveries of 31.3–35.4% with a mean recovery of 33.0 ± 2.1%. Indigenous oocysts detected in these triplicate analyses were detected at concentrations of <2/L, 2/L, and 4/L. In nine analyses, indigenous oocysts were detected five times with concentrations of <2–8/L.

Table 5-12 shows results of *Color*Seed recoveries in secondary effluent samples collected at the three Pennsylvania plants. The *Color*Seed recoveries range from 1.0% (Plant #3) -55.6% (Plant #2). Mean *Color*Seed recoveries for each plant were 26.6 ± 15.3%, 32.6 ± 16.3% and 22.4 ± 13.9% for Plants #1, #2, and #3, respectively. Duplicate analyses on secondary effluent samples collected at Plant#3 in May yielded a mean *Color*Seed recovery of 37.9 ± 7.8%. Ten indigenous oocysts were detected in each 10 L sample. Concentrations of indigenous oocysts detected at Plant #1 were <0.1–3.2/L, <0.1–0.3/L at Plant #2, and <0.1–3.7/L at Plant #3.

Table 5-11. ColorSeed Recoveries and Number of Indigenous Oocysts Detected in Raw Wastewater Influents and Primary Effluents (500-mL samples) from Pennsylvania Plant #3 from November 2002 through July 2003.

Matrix	Month	Turbidity (ntu)	Packed Pellet Volume (mL)	*Color*Seed[1] % Recovery	Number of IO[2] Detected in 500 mL Sample Volume Examined
Raw Influent	November	133	0.4	18.2	1
	December	141	0.75	39.0	0
	January	89.1	0.5	4.0	3
	March	56.3	0.5	28.3	0
	April	58.4	0.4	30.3	1
	May	110	0.5	5.1	3
	May	110	0.5	26.2	4
	May	110	0.5	18.2	2
	June	34.3	0.2	39.0	1
	July	102	0.5	4.0	ND[3]
	Mean *Color*Seed Recovery ± SD			**21.2 ± 13.6**	
Primary Effluent	November	83.5	0.3	26.3	0
	December	39.3	0.3	44.4	0
	January	49.5	0.3	32.3	3
	March	39.4	0.4	35.4	2
	April	41.2	1.0	20.2	0
	May	55.9	0.2	31.3	2
	May	55.9	0.2	32.3	0
	May	55.9	0.2	35.4	1
	June	68.9	0.3	15.2	4
	July	126	0.5	18.2	ND[3]
	Mean *Color*Seed Recovery ± SD			**29.1 ± 9.0**	

1. *Color*Seed spike dose concentration 99 – 100 oocysts.
2. IO, indigenous oocysts.
3. ND, not determined.

Table 5-13 summarizes the occurrence survey results for each of the matrices. Turbidity values for raw influent samples were 34.3–700 ntu. *Color*Seed recoveries were 0–83.8% with a mean recovery of 26.1 ± 17.7%. Indigenous oocysts were detected in 29.5% of raw influent sample analyzed, at concentrations of <2–24/L. In primary effluents *Color*Seed recoveries were 0–78.8% with a mean *Color*Seed recovery of 33.0 ± 17.9%. The turbidity of the primary samples analyzed were 4.08–27 ntu. Indigenous oocyst concentrations were <2–86/L with oocysts detected in 45.8% of these samples. In secondary effluents *Color*Seed recoveries were 0–62.6%. A mean *Color*Seed recovery of 25.0 ± 16.6% was realized for this matrix. Indigenous oocyst concentrations were found to be <0.1–40.8/L in 58.5% of secondary effluents analyzed. In a majority of these samples, 10 L of secondary effluent could be collected on a single Envirochek HV filter. Of 94 samples collected, only six were less than 10 L. Tertiary effluent samples had the highest mean *Color*Seed recovery of 48.8 ± 14.5% with recoveries of 13.1–65.7%. Indigenous oocyst concentrations were <0.008–0.226/L and were detected in 18.8% of tertiary effluent samples analyzed. As a note, tertiary effluent samples were collected at plants in Vermont and North Carolina with indigenous oocysts only being detected in the tertiary effluent from the plant in North Carolina.

Table 5-12. *Color*Seed Recoveries in Secondary Effluent Samples Collected from Three Pennsylvania WWTPs from November 2002 through July 2003.

Plant #	Month	Sample Volume Analyzed (L)	Turbidity (ntu)	Packed Pellet Volume (mL)	*Color*Seed[1] % Recovery	Number of IO[2] Detected
1	November	10	NR[3]	0.2	16.2	12
	December	10	5.92	0.75	36.4	32
	January	10	6.37	0.5	7.1	0
	April	10	1.69	1.0	13.1	13
	May	10	5.09	0.2	34.3	8
	June	10	NR	0.2	28.3	22
	July	10	3.28	0.2	50.6	ND[4]
	Mean *Color*Seed Recovery at Plant #1				**26.6 ± 15.3%**	
2	November	10	NR	0.2	18.2	0
	December	7.0	1.01	0.1	55.6	5
	January	10	1.24	0.2	39.4	2
	March	10	NR	0.5	8.0	0
	April	7.0	0.66	0.2	20.2	0
	May	10	2.68	0.2	45.5	3
	June	10	NR	0.1	44.4	0
	July	10	1.93	0.1	29.2	ND[4]
	Mean *Color*Seed Recovery at Plant #2				**32.6 ± 16.3%**	
3	November	10	NR	0.1	18.2	0
	December	4.2	6.06	0.4	1.0	1
	January	10	2.80	0.3	27.3	37
	March	5.0	8.69	0.5	8.1	0
	April	10	0.75	0.3	11.1	2
	May	10	2.19	0.2	32.3	10
	May	10	1.55	0.3	43.4	10
	June	10	NR	0.2	25.3	29
	July	10	2.18	0.1	35.4	ND[4]
	Mean *Color*Seed Recovery at Plant #3				**22.4 ± 13.9%**	

1. *Color*Seed spike dose concentrations: 99 or 100.
2. IO, indigenous oocysts.
3. NR , not recorded.
4. ND, not determined.

Table 5-13. Summary of *Color*Seed Recoveries, Turbidities, and Number of Samples with Indigenous *Cryptosporidium* Oocysts in Wastewater Samples Collected from May 2002 through July 2003.

Matrix (*n*)	Volume Analyzed (L)	Range of Turbidities (ntu)	Range of *Color*Seed Recoveries (%)	Mean *Color*Seed Recoveries (%)	Concentration of IO[1] (# / L)	Percent of samples positive for IO[1]
Raw Influent (95)	0.5	34.3 – 700	0 - 83.8	26.1 ± 17.7	<2 – 24	29.5
Primary Effluent (84)	0.5	4.08 – 327	0 – 78.8	33.0 ± 17.9	<2 – 86	45.8
Secondary Effluent (94)	4.2 – 22.7	0.75 – 19.3	0 – 62.6	25.0 ± 16.6	<0.1 – 40.8	58.5
Tertiary Effluent (16)	22.7 – 131.3	0.75 – 2.62	13.1 – 65.7	48.8 ± 14.5	<0.008 – 0.226	18.8

1. IO, indigenous oocysts.

Several occurrence surveys have been conducted to monitor for the presence of *Cryptosporidium* spp. oocysts in raw sewage and data indicate a range between 4.1 and 13,700 oocysts/L for the United States and 2.5–800 oocysts/L for the United Kingdom (Bukhari et al., 1997; Madore et al., 1987; Roberson et al., 2000; Parker, 1993). The numbers of oocysts recovered from samples, taken from either a single WWTP or from different WWTPs, demonstrate high variability. In raw wastewater variations in oocyst concentrations are caused by numerous factors, including the levels of infection within the contributing community, the severity of disease among the infected individuals, and the size of the community. In a study by Madore et al. (1987), it was noted that agricultural discharges contributed the highest numbers of oocysts into raw wastewater (≤13,700 oocysts/L). Bukhari et al. (1997) reported oocyst levels of 10–170/L in sewage influents from seven plants located in the England whereas Robertson et al. (2000) reported oocyst levels in <20–400/L at six plants in central Scotland. In another study from the United Kingdom, slaughterhouse effluents were determined to be capable of releasing more than to 149,100 oocysts/L (Parker, 1993). Because effluents from either cattle markets or abattoirs have the potential to release large numbers of oocysts into raw wastewater, the numbers of oocysts present in raw wastewater are also likely to be influenced by contributions from such sources.

Variations in reported oocyst concentrations may be because of the methods used to isolate oocysts from contaminating debris. In most occurrence surveys the concentrates were subjected to isolation and purification techniques that are not specific to *Cryptosporidium.* Density gradient flotation procedures can isolate other biological particles that mask the target organisms during the detection phase of the assay or cross-react with the mAb used to tag *Cryptosporidium* oocysts. Significant numbers of oocysts have been shown to be lost in density gradient centrifugation (Fricker, 1995). In addition, it has been reported that some density gradient procedures selectively isolate viable oocysts (Bukhari & Smith, 1995). McCuin et al. (2000) reported that the IMS procedure does not differentiate viable from nonviable oocysts and can recover both at similar rates. In addition, IMS specifically isolates *Cryptosporidium* from interfering debris, thereby reducing the number of cross reactors present, and with the inclusion of the DAPI stain that targets the sporozoite nucleic DNA, detection of oocysts in environmental matrices is improved.

The incorporation of the *Color*Seed internal control in the current survey was an important tool to evaluate the method performance with each sample. Variability in the wastewater matrices, especially raw influents and primary effluents, were caused by high particle levels, fecal fats, oil and grease, and unknown chemical interferences. This variability creates challenges for recovering *Cryptosporidium.* The internal control may provide evidence of matrix interference as demonstrated on a day when no seeded oocysts were recovered in a secondary effluent sample. It was later determined that dairy waste from a local dairy cooperative had been introduced into the waste stream. Chemicals present in wastewater, such as iron, may inhibit the performance of the IMS kit as noted by Yakub and Stadterman-Knauer (2000). However, optimizing IMS performance in the presence of these interferences was beyond the scope of this research project. In addition, *Color*Seed recoveries may not be indicative of the method's ability to recover environmentally stressed oocysts. Environmentally stressed oocysts may have damaged surface epitopes, making their capture by the IMS difficult, or the oocysts may bind with biological particles rendering the epitopes inaccessible. Some of these interferences were overcome by the addition of detergent and kaolin to the raw influent and primary effluent samples. *Color*Seed recovery data should not be used to determine the number of indigenous oocysts present in wastewater and should only be used to determine method performance in a given sample.

5.3.2 Indigenous Oocyst Concentrations

One objective of this survey was to assess indigenous oocyst concentrations through the wastewater treatment process to determine removal. Figures 5-1 through 5-9 present these data. Data from the Vermont plant are not presented in a figure because oocysts were detected only once at this plant in the raw influent, and never in the secondary or tertiary effluents.

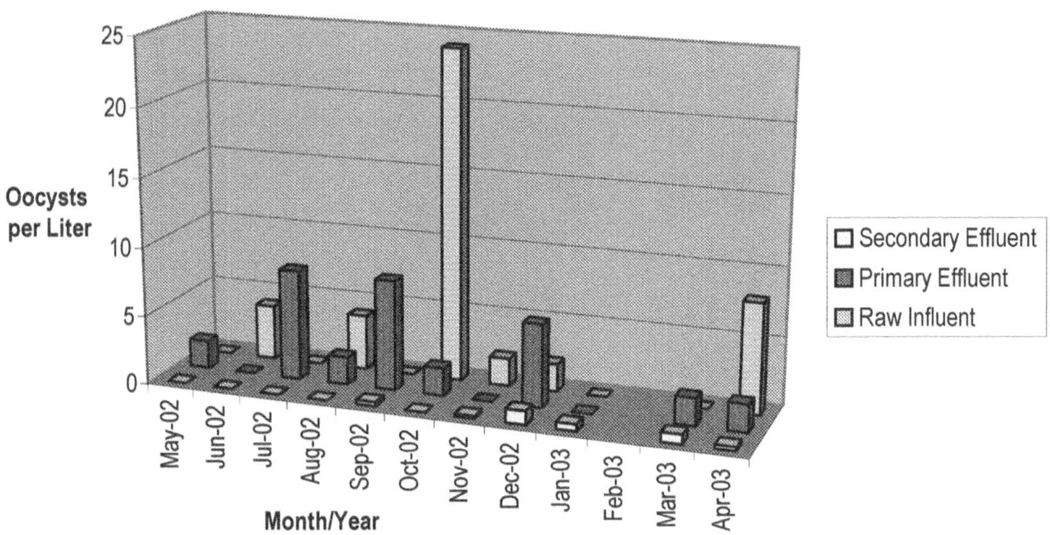

Figure 5-1. Indigenous Oocyst Concentrations at Alabama Plant #1 from May 2002 through April 2003.

Figures 5-1 through 5-3 represent oocyst concentrations detected at the three Alabama WWTPs. Samples were collected from May 2002 through April 2003 with no samples collected in February 2003. Log removals are generally less than 1.0, indicating poor overall removal through the treatment process. In some cases increases in oocyst concentrations from the raw influent to the final influent were noted.

At the California WWTP (Figure 5-4) an increase in oocyst concentrations from the raw influent to the final effluent was observed five of nine sampling events. However, when comparing the log removal from the primary effluent to the final effluent an increase was noted only twice. Oocyst concentrations from samples collected March 2003 showed an increase when the log removal was calculated using the raw influent and final effluent. However, when the primary effluent was used to determine oocyst removal, there was a 1.5 log reduction in oocyst concentrations.

At the Colorado facility (Figure 5-5) log removals were poor when using the raw influent data to calculate removals; however, when using the primary effluent data oocyst removals improved with three datasets to at least 1.0 log and in one case to 2.1 log. At the North Carolina facility (Figure 5-6) oocyst log removals were poor with an increase in oocyst concentrations from the secondary to tertiary effluent noted once. Data from the Pennsylvania plants (Figures 5-7 through 5-9) also showed minimal removals.

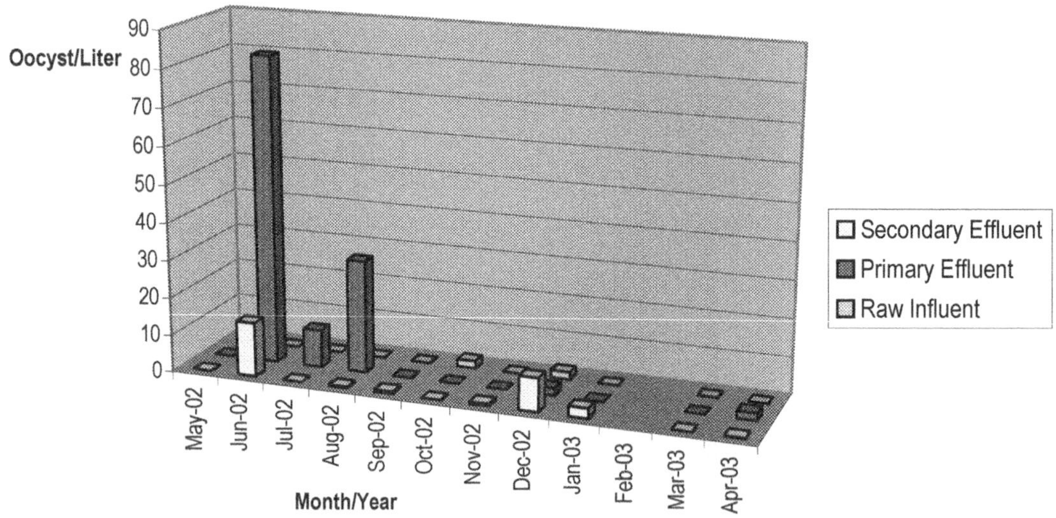

Figure 5-2. Indigenous Oocyst Concentrations at Alabama Plant #2 from May 2002 through April 2003.

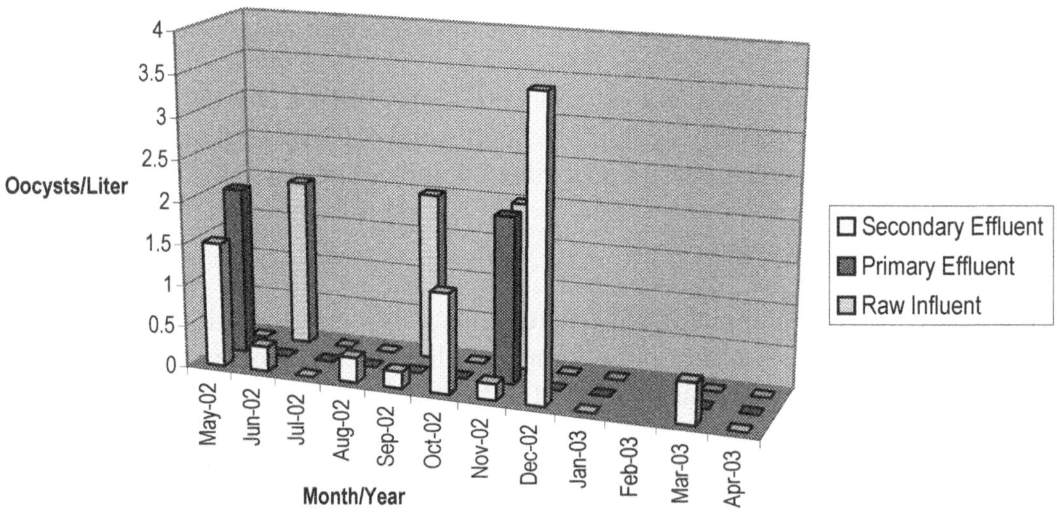

Figure 5-3. Indigenous Oocyst Concentrations at Alabama Plant #3 from May 2002 through April 2003.

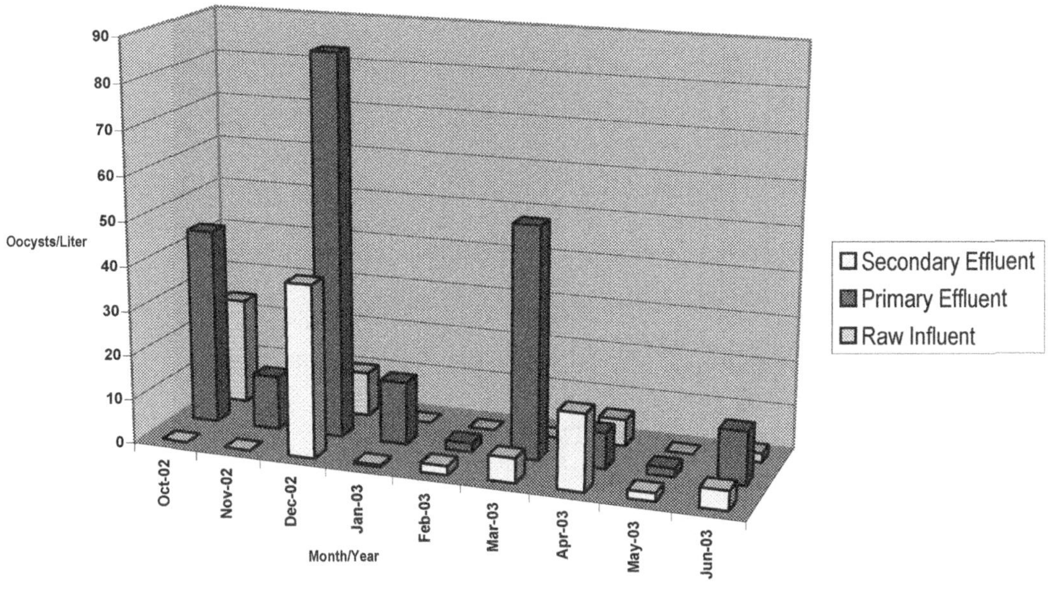

Figure 5-4. Indigenous Oocyst Concentrations at the California Plant from October 2002 through June 2003.

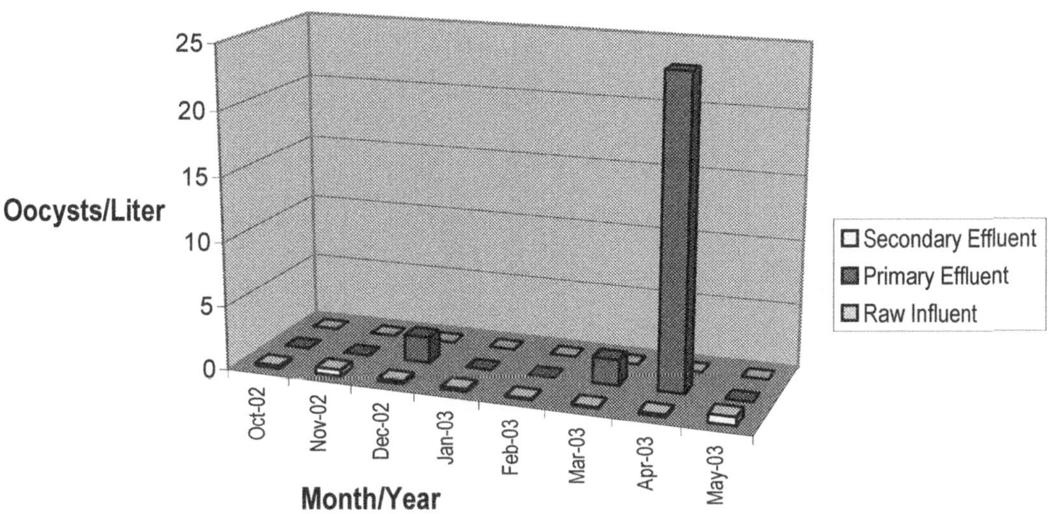

Figure 5-5. Indigenous Oocyst Concentrations at the Colorado Plant from October 2002 through June 2003.

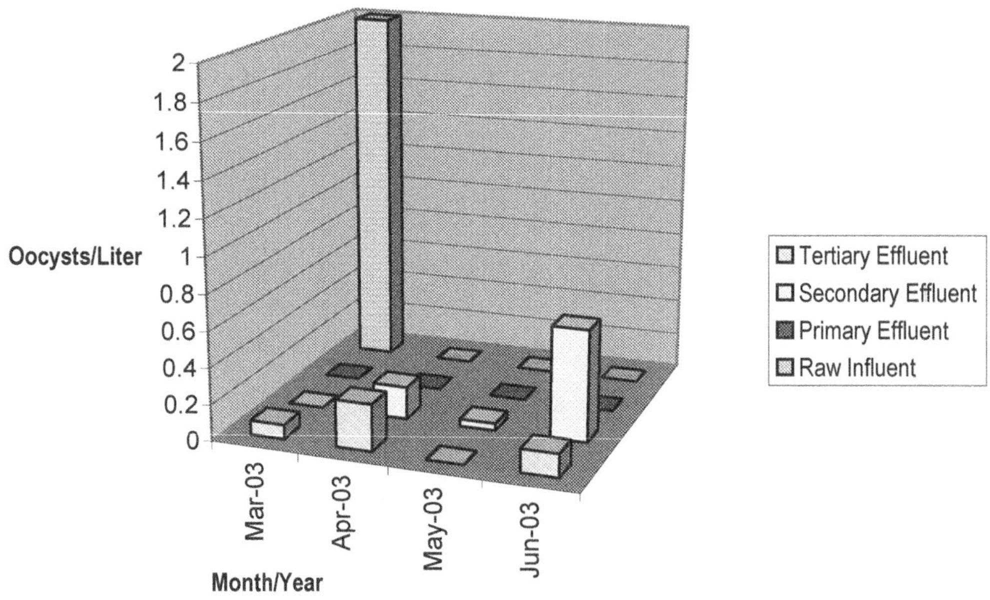

Figure 5-6. Indigenous Oocyst Concentrations at the North Carolina Plant from March 2003 through June 2003.

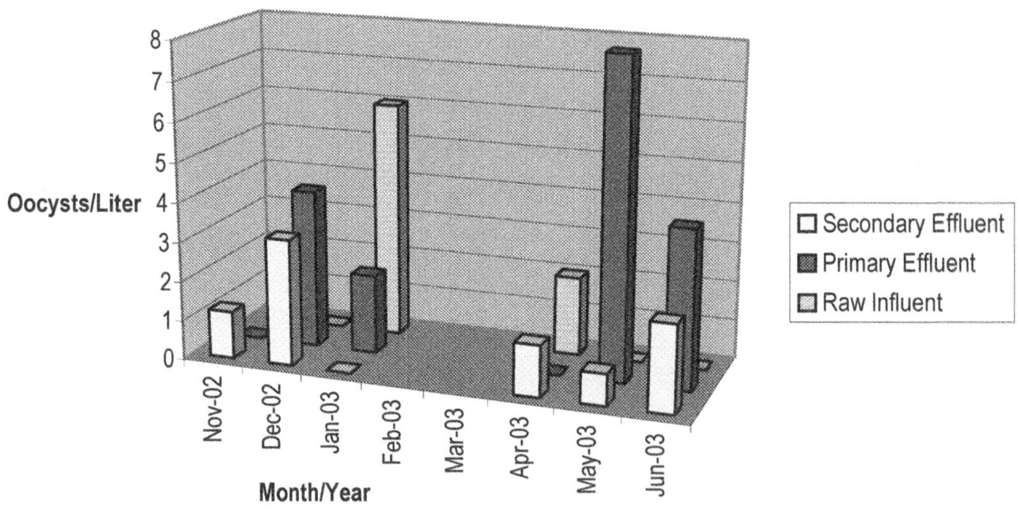

Figure 5-7. Indigenous Oocyst Concentrations at Pennsylvania Plant #1 from November 2002 through June 2003.

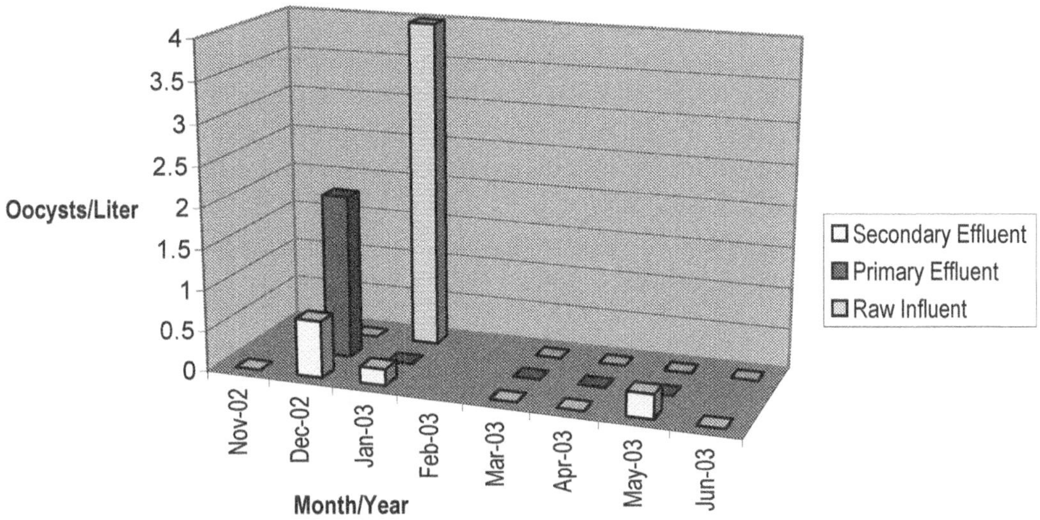

Figure 5-8. Indigenous Oocyst Concentrations at Pennsylvania Plant #2 from November 2002 through June 2003.

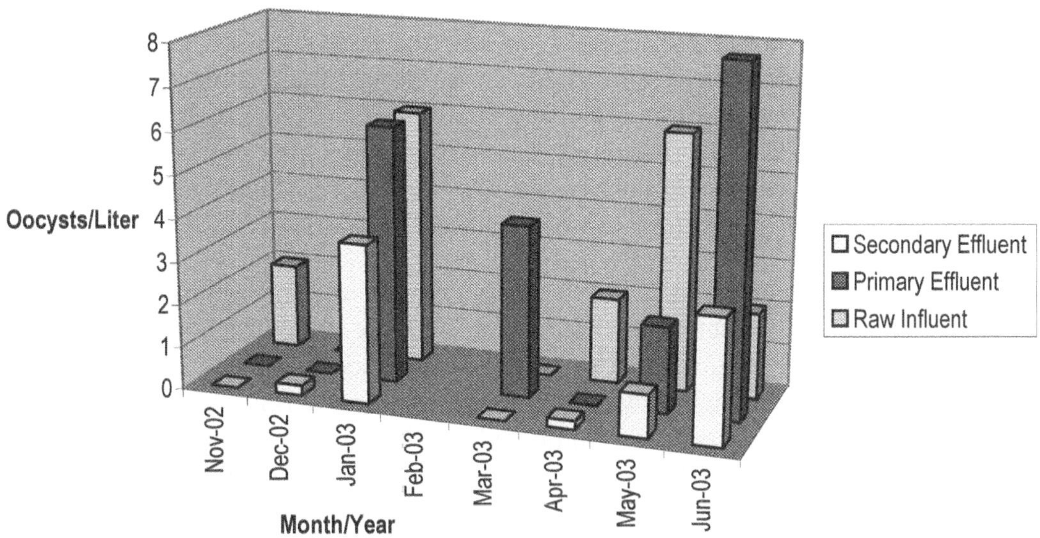

Figure 5-9. Indigenous Oocyst Concentrations at Pennsylvania Plant #3 from November 2002 through June 2003.

A survey conducted in Arizona compared the numbers of *Cryptosporidium* spp. oocysts detected in raw and treated wastewater and for a wastewater treatment works, which used activated sludge, the oocyst removal efficiencies were determined to approach 79% or 0.7 log (Madore et al., 1987). It was also noted that plants using sand filtration and activated sludge released significantly lower numbers of oocysts in their final effluents (i.e., 10 oocysts/L) when compared with plants utilizing activated sludge treatment only (i.e., 1300 oocysts/ L).

In the United Kingdom, Parker (1993) collected six large volume samples (between 100 L and 500 L) of wastewater influent from a municipal treatment works and found oocysts in all six samples at concentrations between 2.5 and 75 oocysts/L. Samples of treated effluents from 10 wastewater works also indicated the presence of oocysts in 35.1% (25/70) of the samples, with concentrations of 0.024–26.5 oocysts/L. In Parker's study, examination of animal slaughterhouse and poultry slaughterhouse effluents indicated that these sources contributed large numbers of oocysts into wastewater influents (92.5–149,100 oocysts/L). For one WWTP, comparison of mean and median oocyst concentrations in influent samples (2.5–75 oocysts/L) and treated effluent samples (0.42–26.5 oocysts/L) indicated oocyst removal efficiencies of 79.6% and 97.7%, respectively.

A Canadian study determined that samples from the five largest population centers in the Yukon contained 26–3022 *Giardia* cysts/L and 0–74 *Cryptosporidium* oocysts/L, and following wastewater treatment, the occurrence of *Cryptosporidium* oocysts exceeded *Giardia* cysts (Roach et al., 1993). These differences in occurrence of *Giardia* and *Cryptosporidium* in raw and treated wastewater may have been caused by higher removal efficiencies of the larger-sized *Giardia* cysts. It is also possible that the treated wastewater samples contained lower levels of interfering debris, which made identification of the smaller-sized *Cryptosporidium* oocysts relatively easier.

Another Canadian study indicated more frequent detection of *Giardia* cysts (73%) than *Cryptosporidium* oocysts (6.1%) in raw wastewater. *Giardia* isolated from raw sewage and treated water samples were subjected to biotyping by isoenzyme analysis and karyotyping by pulsed-field gel electrophoresis and indicated random distribution of these biotypes and karyotypes, which was unrelated to geographic factors (Wallis et al., 1996). A South African study found *Giardia* cysts occurred more frequently than *Cryptosporidium* oocysts in sewage samples, and 30% of the samples contained both organisms. Following wastewater treatment, the proportion of samples positive for both organisms reduced to 25% (Kfir et al., 1994).

To assess the removal efficiency in WWTPs, daily effluent and influent samples from six WWTPs in central Scotland were collected for three years, with the aid of 24-hour samplers (Robertson et al., 2000). To ensure that effluent samples corresponded with the influent samples, the former were collected after a time delay, which corresponded to the holding time of the treatment works. Data from these investigations indicated that low numbers of *Cryptosporidium* spp. oocysts were detected in both the influent (<10–160/L) and effluent samples (<5–85/L), and removal efficiencies were difficult to assess. However, in samples where sufficient numbers of oocysts were recovered for calculation of removal efficiency, wide variations were noted within each treatment works.

In a U.S. study using grab samples of wastewater influent and primary/secondary effluents, only 1 log reduction in *Giardia* cysts and *Cryptosporidium* oocysts was noted following primary treatment (Mayer & Palmer, 1996). Secondary treatment increased *Giardia* cysts removal by a further 2 log, whereas *Cryptosporidium* oocysts removal was increased only by a further 1 log. In this study, the authors noted that although levels of organisms were reduced across treatment they were eliminated and hence were released into the environment where they could potentially remain viable for months.

A study in Spain simulated the activated sludge treatment process in a laboratory by preparing a 3:1 ratio of activated sludge to oocyst seeded primary clarifier effluent. Air was bubbled through the mixture for 3 h and flocculated particles were allowed to sediment (35–40 min) and the supernatant was concentrated to 500 mL. This procedure removed 83–84% of the spiked oocysts, and when aliquots of wastewater treated in this manner were inoculated into mice, the oocysts were shown to retain their infectivity (Villacorta et al., 1992).

5.3.3 Oocyst Removal Results

A model of an activated sludge treatment plant (flow rate of 17 mL/min and retention time of 6 h) was used to demonstrate oocyst removal efficiencies of 83.4% and 90.7% following primary and secondary sedimentation, respectively (Stadterman et al., 1994). To evaluate *Cryptosporidium* spp. oocyst removal efficiencies during various wastewater treatment processes, raw sewage, activated sludge (retention time 2.8 h), trickling filter (retention time 3.5 h), and biodisc (retention time 4.8 h) effluents were seeded with oocysts and settled for 2 h or for their respective retention times. When compared with other sedimentation processes that were evaluated, activated sludge yielded the highest oocyst removal efficiency, 92% or 1.1 $\log_{10}$ (Stadterman et al., 1995).

In this survey the low numbers of oocysts detected in the influent and effluent samples hindered our attempts to make valid statements about removal efficiencies for each of the WWTPs. However, levels of removals determined in this study are similar to those reported by other researchers. In previous studies, methods for the isolation of *Cryptosporidium* oocysts in wastewater have relied on nonselective isolation procedures such as density gradient flotation, which can recovery algae and other biological particles. These particles may cross react with the mAbs or occlude oocysts, preventing proper identification. In addition, identification of oocysts relied primarily on fluorescence pattern, size, and shape. Before the late 1990s, very few researchers included the use of DAPI staining or Nomarski DIC optics for visualization of internal structures. Using the older methods (ICR, SCA), the inability to rule out objects similar to *Cryptosporidium* oocysts based on size, shape, and fluorescence patterns alone may have led to overestimations of the true numbers in samples.

Another problem with the early methods was the analysis of small subsamples, where counts were extrapolated to the total volume sampled. Analysis of subsamples with extrapolation to much larger sampled total volumes (e.g., 100 L) has been shown to lead to large errors in determination of true pathogen density (Parkhurst & Stein, 1998; Atherholt & Korn, 1999). This aspect of the early analysis has led to data that fall generally into two broad categories—either nondetects or very high reported levels based on low analyzed sample volume. For example, from a 100 L sample, a portion equivalent to 2.5 L is actually examined microscopically because of method limitations. If one oocyst is observed, then the reported value is 40 oocysts/100 L, assuming incorrectly that oocysts are evenly distributed in a sample. The nondetect data are equally absurd. Using the same example, if no oocysts were detected in the 2.5 L equivalent volume, the count is reported as <40 oocysts/100 L. This means the count could be 39 or 0 or any number in between. These methods were so poor that actual count data as well as nondetect data are unreliable (Allen et al., 2000). A combination of these methods errors may have led researchers to report erroneous results.

In this study we had the opportunity to use IMS to recover oocysts from the wastewater matrices. The IMS technique offers selective isolation resulting in reduced levels of interfering debris. Incorporating an additional wash of the captured bead-oocyst complex before disassociation also reduces these interferences. Although some nontarget organisms may be captured with IMS, the use of the additional fluorogenic stain, DAPI, aids in the identification of recovered oocysts. Incorporating *Color*Seed in the analysis of every sample offers additional quality assurance/quality control, adding validity to the performance of the method in a given sample. In addition, poor *Color*Seed recoveries may indicate the presence of matrix interferences, which may impact the performance of the IMS protocol to recover indigenous oocysts. In all cases, the volume analyzed was the volume used when reporting oocyst recovered. In the presentations comparing the recoveries in each type of sample, the raw and primary effluent data were doubled (presented as per liter rather than per 500 milliliters) and the secondary and tertiary effluents were standardized and reported per liter, so that direct comparisons could be made using a standard volume. This was necessary to develop the log removal data.

This occurrence survey used the most current tools to recover *C. parvum* oocysts from wastewater. These technologies present us with an opportunity to provide QA/QC data on every sample analyzed while offering the most selective oocyst isolation procedure available to date. These data show that oocysts are present in wastewater matrices and more importantly are present in effluents being discharged from the plant. The questions that remain to be answered: are these oocysts viable and more importantly, able to cause infection.

CHAPTER 6.0

AN ASSAY TO DETERMINE INFECTIVITY OF *CRYPTOSPORIDIUM* OOCYSTS IN WASTEWATER

6.1 Introduction

Conventional methods for the detection of *Cryptosporidium* in environmental matrices are labor intensive. Recovered particles are stained with a fluorescent mAb and detected using epifluorescence microscopy. Although criteria exists to identify oocysts recovered in environmental matrices, the microscopy assay is subjective and requires dozens of hours of training to become competent at identifying oocysts. Objects that are the correct size, shape, and exhibit the appropriate fluorescence pattern are confirmed by examination of the DAPI staining characteristics, followed by visualization of the internal structures using DIC microscopy. A limitation of the assay is the nonspecific binding of the antibody to nontarget organisms. Some of these nontarget organisms can appear to be oocysts based on the identification criteria used. The reagents used to detect oocysts do not discriminate between *C. parvum* and *C. hominis*, which are human pathogens, and other *Cryptosporidium* species. In addition, the method cannot determine if recovered oocysts are viable, and more important, infectious.

Over the past decade researchers have spent considerable effort in developing cell culture and molecular techniques to detect *Cryptosporidium* in environmental matrices. The advantages in using these techniques to detect *Cryptosporidium* in environmental waters are their potential for greater sensitivity and ability to distinguish between species, if appropriate primers, targeting a variety of genes or genomic fragments, are chosen. In addition, many samples can be analyzed rapidly at relatively low costs. *In vitro* cell culture techniques, such as *in situ* hybridization or foci detection assays can provide information on the relative levels of infectious oocysts detected in the each sample, whereas molecular techniques such as polymerase chain reaction (PCR) and reverse transcription (RT)–PCR can provide information regarding species and relative quantification of oocysts detected. Another molecular technique currently being developed is real-time PCR that uses a fluorescent probe during PCR cycling to label target amplicons, negating the need for post-PCR processing.

More than 20 cell lines have been reported capable of supporting *Cryptosporidium* infection assays. Some of these cell lines have shown high sensitivities to infection with low levels of *Cryptosporidium* oocysts. Although Caco-2 (human colonic adenocarcinoma) cells will differentiate once they reach confluency, forming brush border microvilli and thereby representing a good model of the human intestine, HCT-8 (human ileocaecal adenocarcinoma) cells have become the cell line most widely used by researchers. This is most likely because HCT-8 cells are permissive to infection by low numbers of *Cryptosporidium* oocysts and can support infections by both genotypes (Type 1, human; Type 2, human and animal). Other reasons HCT-8 cells are preferred is they are easy to maintain, they grow relatively quickly and are robust. Media formulations optimized for cell growth do not always correlate with increased parasite development. Upton et al. (1995) reported that supplementing growth media with vitamins increased *Cryptosporidium* growth in cell culture tenfold when compared to media without the added supplements. Fetal bovine serum (FBS) is added to most media formulations, at concentrations of 5–20%, to improve parasite development, but some researchers have suggested that FBS contains inhibitory substances to parasite growth (Arrowood et al., 1994). Some researchers have suggested that pretreatment of sample concentrates before inoculation onto confluent monolayers with acid, 10% bleach, or bile salts improve infection rates by inducing oocysts to excyst releasing the sporozoites to infect host cells (Upton 1994 a, b; Upton 1995, DiGiovanni et al., 1999; Slifko et al., 1999; LeChevallier et al., 2003; Gennaccaro et al., 2003; Quintero-Betancourt et al., 2003). However, Rochelle et al. (1999) suggested that pretreatment of environmentally stressed oocysts may have deleterious effects on oocyst viability. In addition, the authors suggest that synergistic effects may occur when oocysts are pretreated for cell culture infection after exposure to disinfectants in inactivation studies.

Numerous papers have been published in the past decade on the usefulness of PCR to detect and identify species of *Cryptosporidium* found in environmental samples. These studies have used numerous primer sets targeting a variety of genes or genomic fragments, such as the *Cryptosporidium* oocyst wall protein gene, 18S ribosomal RNA (rRNA), β-tubulin messenger RNA (mRNA), small subunit rRNA, repetitive oocyst proteins (873 bp, and 358 bp) and the heat shock protein (*hsp*) 70 gene. Detection sensitivities are varied for these primers and range from 1 to 1000 oocysts. In comparison trials of four reported primers sets for *Cryptosporidium*, Rochelle et al. (1997a) was able to detect 1–10 *C. parvum* oocysts in highly purified preparations and 5–50 oocysts in seeded environmental matrices. In this study the authors reported the LAX primer set, LAX 469F and LAX 869R, was the most sensitive (1 oocyst) and was specific for *C. parvum,* whereas other sets tested were not specific to *C. parvum*. Stinear et al. (1996) developed an RT-PCR assay, which was able to detect one viable oocyst spiked into concentrated environmental water samples when combined with Southern hybridization. This RT-PCR assay was developed to detect mRNA from the *C. parvum hsp* 70 gene (590 bp), induced by exposure to elevated temperatures. The authors targeted this protein because it is produced only by viable oocysts and is produced in large numbers by stressed oocysts. Other researchers have investigated PCR combined with restriction fragment length polymorphism (RFLP) analysis to differentiate *Cryptosporidium* species in environmental matrices (Elwin et al., 2001; Xiao et al., 2001; Guyot et al., 2002; Reed et al., 2002; Sturbaum et al., 2002; Nichols et al., 2003; Wu et al., 2003). Although some investigators may be interested in the different *Cryptosporidium* species present in environmental matrices, it is a primary objective of the water industry to be able to detect viable and infectious oocysts that pose a potential risk to humans. For this reason, cell culture methods combined with molecular techniques have been developed to determine if infectious oocysts are present in these matrices.

Rochelle et al. (1999) have demonstrated that *Cryptosporidium* oocysts recovered using IMS are able to retain their infectivity. Because the methods developed in this research project use IMS to recover *Cryptosporidium* oocysts from the wastewater matrix it seemed appropriate to determine infectivity of recovered oocysts from wastewater matrices incorporating cell culture techniques using the IMS concentrates.

6.2 Material and Methods

6.2.1 CC-QSD Analysis

6.2.1.1 Sample Processing of Raw Sewage Influents and Primary Effluents

One-liter samples of raw sewage influent or primary effluent were collected from four WWTPs. Each sample was shaken vigorously to resuspend large particles and solids present and a 500-mL subsample was transferred to a 550 mL sample bottle. A volume of Tween 80 was added to the sample to a final concentration of 1% and shaken to evenly distribute. The sample was concentrated by centrifugation (1500 x *g*; 15 min) and the supernatant was aspirated to above the packed pellet material. The final sample concentrate was resuspended and transferred to a Leighton tube and autoclaved at 121°C, 16 psi for 15 min. Autoclaving the samples inactivated infectious indigenous oocysts that were potentially present so that infectious oocysts detected in the CC-QSD assay would be from seeded oocysts only. The sample was cooled and seeded with one vial *Color*Seed and 579 ± 45 viable oocysts. Seeded oocysts were isolated from the contaminating debris using IMS in the presence of 0.75 g kaolin. After the required incubation period, the oocysts were isolated as described in Section 2.2.4.

6.2.1.2 Sample Processing of Secondary and Tertiary Effluents

Secondary effluent samples were collected from WWTPs. The filter was attached to the source and up to 23 L of sample passed through the filter at a flow rate of 2 L/min. Retained particles were eluted and isolated from the filter as previously described. The sample concentrate was transferred to a Leighton tube and autoclaved at 121°C, 16 psi for 15 min. Sample concentrates were autoclaved to inactivate indigenous oocysts so that infections occurring in the assay could only be derived from seeded viable oocysts. The autoclaved concentrate was cooled and seeded with one vial of *Color*Seed and 579 ± 45 viable oocysts. Seeded oocysts were isolated from the contaminating debris using IMS as described in Section 2.2.4.

6.2.1.3 Processing of IMS Concentrates

An additional rinse of the bead-oocyst complex was performed in the Leighton tube with 10 mL PBS. The rinsed complex was isolated using the MPC-1, resuspended in 1 mL PBS and transferred to a 1.5 mL Eppendorf tube. The Leighton tube was rinsed with an additional 0.5 mL of PBS, which was added to the Eppendorf tube. The bead-oocyst complex was isolated from the rinse solution using the MPC-M. The beads were rinsed in another 1 mL aliquot of PBS and then suspended in 100 μL Hank's Balanced salts solution (HBSS). Fifty microliters of the resulting IMS concentrate was transferred to a 1.5 mL Eppendorf tube and sent to Texas A&M University (TAMU) for cell culture analysis using the CC-QSD method.

6.2.1.4 St. Albans Secondary Effluent Samples

To determine method sensitivity, five replicate 6 L samples of secondary effluent from the St. Albans WWTP were collected on HV filters, eluted, and concentrated as described above. Note that no oocysts or *Color*Seed were added to the filters before filtration. The sample concentrates were autoclaved and processed by IMS as described above with the final bead-oocyst complexes each resuspended in 100 μL of HBSS. Viable *C. parvum* oocysts were added to each sample in 100 μL aliquots. Oocyst concentrations were determined using 10-well counts, as described in Section 2.2.1.1. Oocyst concentrations were determined to be: 0, 8.5 ± 3.3, 26.5 ± 4.7, 49.7 ± 8.2, and 98.0 ± 10.7. Each vial was sent to TAMU for analysis within 48 h of sample collection.

6.2.1.5 Analysis of Wastewater Samples Using CC-QSD

Cryptosporidium oocysts recovered from wastewater samples suspended in HBSS were processed for CC-QSD as previously described (LeChevallier et al., 2003). Briefly, IMS concentrates were prepared for HCT-8 cell culture by suspension in acidified HBSS (pH 2.0) containing 1% trypsin (AHBSS/1% trypsin), followed by incubation at 37°C for 1 h with vortexing every 15 min. Beads were removed from the acidified suspension by application of a magnetic field. Resulting supernatants containing the dissociated oocysts were transferred to new microfuge tubes. To ensure recovery of oocysts, a second wash of the IMS beads with 100 µl of AHBSS/1% trypsin was performed. Growth Medium (RPMI 1640) containing 10% FBS (150 µl) was added to each sample and then centrifuged at 15,800 x *g* for 2 min at room temperature. Samples were carefully aspirated to 50 µL. A second wash of the samples was performed to ensure removal of trypsin by adding 500 µl of Growth Medium. Samples were centrifuged and then aspirated to 20 µl. Each sample was then resuspended in 80 µl of prewarmed (37°C) Growth Medium containing 0.05% trypsin. Just before inoculation of monolayers, 500 µl of Maintenance Medium were removed. Monolayers were inoculated and then incubated at 37°C in a 5% CO_2 humidified incubator for 72 h before DNA extraction. Positive controls consisted of AHBSS/1% trypsin-treated, viable *C. parvum* oocysts were 10–289 oocysts per monolayer. Negative controls were uninoculated monolayers. Intact oocyst no infection controls (IONICs) consisted of intact viable oocysts (approximately 10–289 oocysts per monolayer) added to the cell monolayers immediately before DNA extraction. IONICs were used to evaluate potential CC-QSD background from intact oocysts remaining on monolayers. Samples were applied to cell monolayers within 7 days of sample collection.

After incubation, the cell monolayers were washed five times with 200 µL of PBS to remove unexcysted oocysts. DNA was purified from HCT-8 cells using the QIAamp DNA mini kit (Qiagen, Inc., Valencia, Calif.) and a modified extraction protocol. Modifications included incubation of the cell monolayers with lysis buffer for 15 min at 70–75°C, and elution of DNA from the QIAamp DNA columns using 50 µl of 0.01X TE pH 8.0 at 70–75°C for 5 min. QSD reaction conditions, primers, and TaqMan probe were as previously described (DiGiovanni et al., 2002; LeChevallier et al., 2003), with the exception that PCR product carryover prevention was included through the use of deoxyuridine triphosphate and uracil DNA glycosylase.

6.2.2 FDM Analysis

6.2.2.1 Sample Processing for FDM

Six individual 10-L volumes of secondary effluent from two treatment plants were collected on individual Envirochek HV capsule filters and processed as previously described. Resulting bead-oocyst complexes were resuspended in 100 µL HBSS. Duplicate IMS concentrates from each facility were analyzed unseeded. In addition, duplicate IMS concentrates from each facility were seeded with 100 µL samples of working *Cryptosporidium* suspensions containing approximately 25 or 100 oocysts. *Cryptosporidium* oocysts used for this study were obtained from the Sterling Parasitology Research Laboratory at the University of Arizona (Tucson) and were three weeks post-animal shedding. Suspensions of oocysts used for seeding were prepared and enumerated as previously described (Chapter 2.0, Section 2.2.1.1). Table 6-1 presents the number of replicates and seeding levels used in this study. In addition, a suspension of the *Cryptosporidium* oocysts used was provided to TAMU for controls in the FDM.

Table 6-1. Seeding Levels for Samples Analyzed by FDM.

WWTP	*C. parvum* Oocyst Concentration	No. of Replicates
1	0	2
	22.2 ± 2.5	2
	85.1 ± 6.5	2
2	0	2
	22.2 ± 2.5	2
	85.1 ± 6.5	2

One half (100 μL) of each sample containing resuspended beads was transferred to individual clean 1.5 mL Eppendorf tube and sent to TAMU for detection of infectious oocysts using FDM within 48 h of sample collection. Samples were applied to cell monolayers the day they were received. The remainder of the sample was applied to a well slide, dried, and stained with FITC-mAb and DAPI. Recovered oocysts were enumerated using epifluorescence microscopy.

6.2.2.2 FDM for Determining the Presence of Infectious Oocysts in Wastewater

The IMS concentrates were processed for inoculation on to HCT-8 monolayers using the same protocol described above for CC-QSD except that the final concentrate was resuspended in 500 μL of growth medium containing 0.05% trypsin. Samples were inoculated onto single HCT-8 monolayers cultivated in 8-well chamber slides. Positive and negative controls consisted of suspensions containing 100 oocysts and 0 oocysts, respectively. Slides were incubated in 5% CO_2 atmosphere at 37°C for 72 h. After the incubation period, the growth medium was removed from the wells and the infected monolayers were washed to remove unattached oocysts. Wells were fixed with absolute methanol for 8 min and stained with fluorescein-labeled polyclonal rat immunoglobulin G made to sporozoites of *C. parvum* (Sporo-Glo, Waterborne, Inc., New Orleans, LA). Chamber slides were examined using epifluorescence microscopy and infectious foci were enumerated.

6.3 Results and Discussion

6.3.1 CC-QSD Results

CC-QSD is an assay that is theoretically able to determine the numbers of oocysts present in a sample based on the quantification of target DNA during the PCR assay. During the early exponential phase of the amplification reaction, template DNA and the amplified products are directly proportional. A TaqMan fluorogenic detection system is used to measure these products during each PCR cycle. The hybridization probe is labeled with two fluorogenic stains. As the DNA polymerase degrades the probe during the elongation step an increase in fluorescence is measured. The threshold cycle (or Ct value) is defined as the point during cycling when the amplification of the PCR product is first detected. Therefore the higher the starting copy of the DNA target the sooner an increase in fluorescence is observed. In other words, fewer cycles are required to detect the target DNA as the number of oocysts is increased.

Seed stock oocysts (~289 oocysts/100 μL) were used to determine the CC-QSD baseline cell culture infection rates. The oocyst infection rates of the monolayers were determined to be 5% and 8% using 25 and 10 oocysts per monolayer, respectively. Results for the IONIC samples revealed a 6% infection rate for 289 oocysts; the IONIC sample using 10 oocysts tested negative.

The infection rates obtained in these trials are similar to those reported by Rochelle et al. (2001). On the basis of these CC-QSD values, percent recoveries for the seeded environmental samples were calculated. Most seeded environmental samples tested positive. However, CC-QSD infectious oocyst recovery rates were from 30% to more than 200,000% (see Table 6-2). In addition, no amplification of the seeded oocysts were detected in the St. Albans samples. One possible explanation is that there was some type of matrix effect that resulted in unusual and variable amplification of the seeded wastewater samples. In this study, the wastewater concentrates were autoclaved, which may have affected the chemical-physical properties of the samples. Another possible explanation is that the TaqMan probe fluorescence was affected by the sample matrices; however, this would likely be expressed as a decrease in fluorescence rather than an increase. However, Fontaine and Guillot (2002) noted that amplification of nontarget sequences could occur when the free magnesium concentration is not optimized resulting in an increased fluorescence signal. The amplification plot for the samples are presented in Figure 6-1 and appeared normal and robust. *Color*Seed recoveries obtained in these trials were similar to those obtained in previous trials for each matrix tested (raw influent, primary, secondary, and tertiary effluents).

6.3.2 FDM Results in Secondary Effluents

Table 6-3 presents results from the FDM trials for determining the number of infectious oocysts in secondary effluent samples. Samples collected at the wastewater facility in California show the presence of infectious *Cryptosporidium* oocysts at levels of 2–6 (samples 1–6). The unfortunate aspect of these trials is that the seeded oocysts had a low level of infectious oocysts, as demonstrated from the controls analyzed. At a seeding level of ~200 oocysts/100µL, only one infectious focus was detected in two replicates. The infectious foci detected in seeded replicates support these findings (see Table 6-3). The results from the FITC-mAb counts show a recovery of 139% at a seeding level of approximately 25 oocysts. The high recovery rate may in fact be caused by higher numbers of indigenous oocysts present in that subsample as compared to those levels detected in the unseeded subsamples analyzed. Figures 6-2 though 6-4 show infectious foci detected in the secondary effluent samples.

The methods evaluated to determine the presence of infectious oocysts in wastewater show very different results. Cell culture combined with PCR (CC-PCR) provides a high degree of specificity for the detection of infectious *Cryptosporidium* oocysts in water samples but is only a presence/absence assay. By incorporating a real-time PCR assay that employs a high-performance DNA polymerase for nucleic acid amplification and a 5' nuclease assay using a fluorogenic (TaqMan) DNA probe, quantitative results can be determined as demonstrated from seeded environmental water samples (LeChevallier et al., 2003). In that study, the recovery efficiency averaged 58.5% for CC-PCR samples and 72% for Method 1623. These differences were not significantly different ($P = 0.06$). In the present study, the CC-QSD method appears to have several problems in detecting infectious oocysts in seeded IMS concentrates derived from wastewater despite attempts to remove potential interferences by autoclaving the sample concentrates before IMS and performing additional rinses of the bead-oocyst complex before seeding with purified suspensions of *C. parvum*. Although analysis of the controls resulted in acceptable results, data from analysis of the wastewater samples yielded results that indicate

Table 6-2. CC-QSD Results from Wastewater Matrices from Five WWTPs.

Plant Location	Wastewater Matrix	Result (% recovery based on 289-oocyst baseline)	*Color*Seed/FA Recoveries (%)
California	Raw	30	55.6/63.6
	Primary	22624	37.4/66.7
	Secondary	98	42.4/69.8
North Carolina	Raw	No amplification	29.3/35.9
	Primary	41192	19.2/40.4
	Secondary	8163	35.4/64.9
	Tertiary	305	26.2/72.2
Colorado	Raw	518	11.1/15.5
	Primary	661	20.2/27.3
	Secondary	No amplification	16.2/30.1
Pennsylvania	Raw	26885	23.2/52.5
	Primary	893	26.2/49.1
	Secondary	200010	29.2/47.3
Vermont	Secondary - Blank	No amplification	Not Applicable
	Secondary – 100 oocysts	No amplification	Not Applicable
	Secondary – 50 oocysts	No amplification	Not Applicable
	Secondary – 25 oocysts	No amplification	Not Applicable
	Secondary – 10 oocysts	No amplification	Not Applicable
	289 oocyst baseline	100%	Not Applicable
	25 oocyst baseline	5%	Not Applicable
	10 oocyst baseline	8%	Not Applicable

QA/QC

Sample	Result
CC-QSD positive control	Positive
CC-QSD 289 intact oocyst no infection negative control	6% of 289 oocyst infection baseline
CC-QSD 10 intact oocyst no infection negative control	Negative
CC-QSD method blank (uninoculated cell monolayer)	Negative
PCR no template negative control	Negative
PCR positive control	Positive

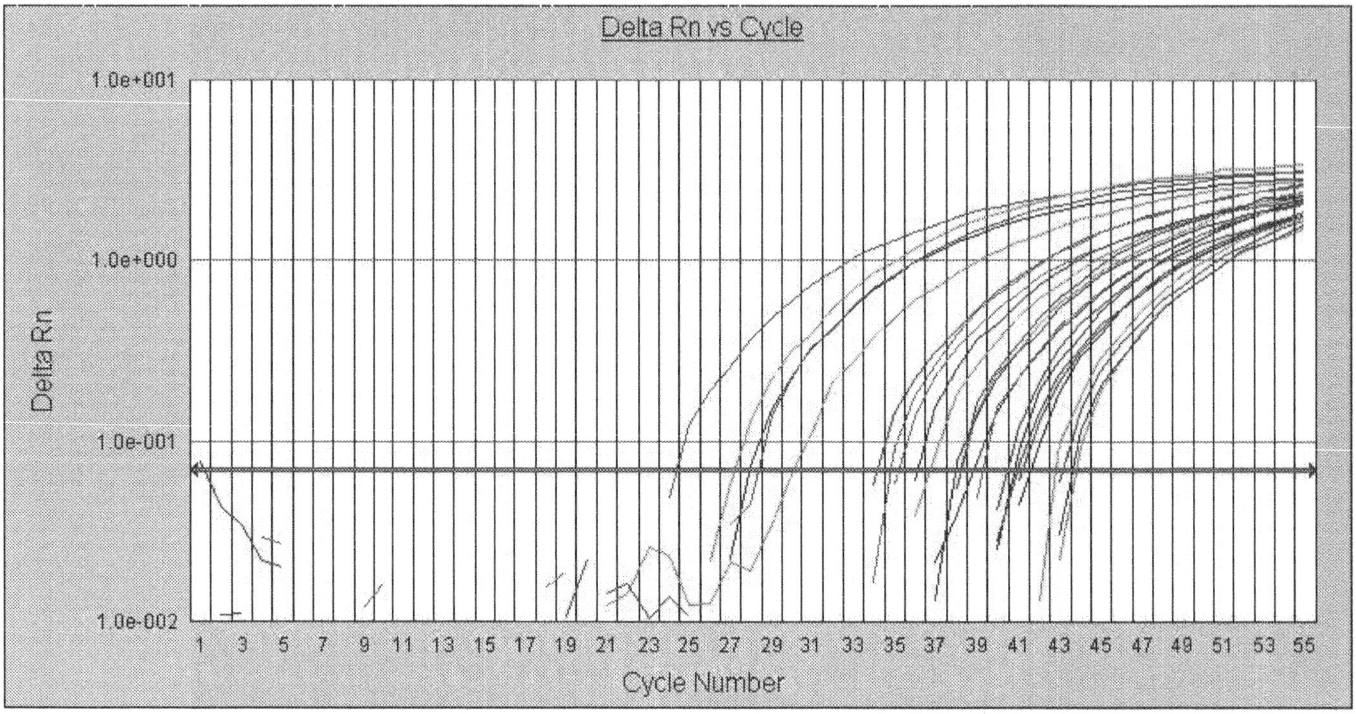

Figure 6-1. CC-QSD Amplification Plot. Note: X axis is PCR cycle number and Y axis is change in normalized fluorescence reporter (a measurement of PCR amplification). Each curve represents amplification of individual samples. The heavy bar is the threshold of detection and fractional number of PCR cycles at which sample fluorescence crosses threshold (cycle threshold, or Ct, value). Ct values are used to quantify infection based on a composite CC-QSD standard curve. Amplification curves appear normal and robust. Negative controls are below Y axis range.

Table 6-3. FDM Results for Secondary Effluent Samples Seeded with *C. parvum* Oocysts.

Sample Location (Sample ID)	Number of *C. parvum* Oocysts Seeded into Sample	Result	Number of Foci Detected (Estimated Number of Infectious Oocysts)	FA[1] Counts of Seeded *C. parvum* oocysts (Percent Recovery)
California (1)	0	Positive	5	41
(2)	0	Positive	4	34
(3)	22.25 ± 2.5	Positive	5	83 (139)
(4)	22.25 ± 2.5	Positive	2	LA[2]
(5)	85.1 ± 6.5	Positive	5	41 (33.4)
(6)	85.1 ± 6.5	Positive	6	63 (51.4)
Pennsylvania (7)	0	Negative	0	1
(8)	0	Negative	0	0
(9)	22.25 ± 2.5	Negative	0	20 (87.9)
(10)	22.25 ± 2.5	Negative	0	18 (79.1)
(11)	85.1 ± 6.5	Negative	0	94 (110)
(12)	85.1 ± 6.5	Negative	0	97 ((114)
Control	~50/100µL	Negative	0	Not Applicable
Control	~50/100 µL	Negative	0	Not Applicable
Control	~200/100 µL	Negative	0	Not Applicable
Control	~200/100 µL	Positive	1	Not Applicable

1. FA, fluorescent antibodies.
2. LA, laboratory accident – lost sample.

Figure 6-2. Focus of *Cryptosporidium* Cell Culture Infection for Sample 1.

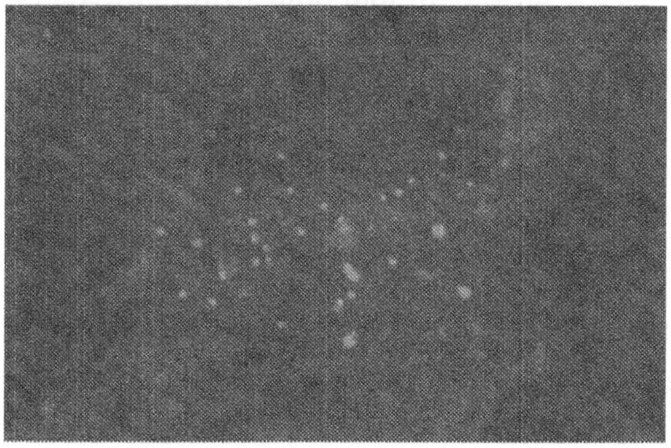

Figure 6-3 Focus of *Cryptosporidium* Cell Culture Infection for Sample 3.

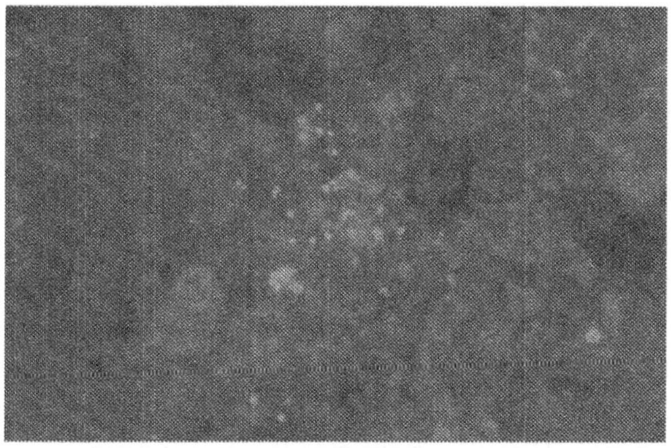

Figure 6-4 Focus of *Cryptosporidium* Cell Culture Infection for Sample 6.

amplification of nontarget sequences. On the basis of these results, we were reluctant to use this method in a second round of testing and sought a different approach for detecting infectious oocysts in these complex matrices.

Researchers using FDM for determining the numbers of infectious oocysts in environmental waters have reported that this method can make these determinations in a range of water matrices (Gennaccaro et al., 2003; Quintero-Betancourt et al., 2003; Slifko et al., 1997).

The matrices tested included DI, high-salt and high-sugar solutions, raw wastewater influents, secondary effluents, and reclaimed water. Gennaccaro et al. (2003) detected infectious *C. parvum* oocysts in 40% of reclaimed effluents and 33% of raw influents tested. Reported concentrations of oocysts were <1.0/100L in postfiltration secondary effluent, 26,300/100L in raw influent based on FA counts, <1/100 L in final reclaim water, 5065/100L in raw influent based on FDM-MPN results. These reported concentrations for both total and FDM/MPN counts far exceed the concentrations reported in the current study. Reported concentrations of oocysts by Gennaccaro et al. (2003) were determined from low numbers observed in small volumes and then extrapolated to 100 L. Allen et al. (2000) observed that reported levels of oocysts detected in source water from ICR data fell into two broad categories based on this approach to data reporting, nondetects and high reported concentrations, and these data are unreliable. They argued that biological particles, such as oocysts, are not evenly distributed in samples and if one oocyst is detected in a small volume, extrapolating the results to a large volume yields incorrect information about actual numbers present. They continue the argument that data with nondetects leads to equally incorrect assumptions about the concentrations of oocysts in these matrices. A value of <400/100 L could mean that there are 399, 15, or 0 oocysts present in 100 L. Another issue with these data is that antibodies used to detect *Cryptosporidium* oocysts in environmental matrices do cross-react with other *Cryptosporidium* sp. as well as algal cells and other nontarget organisms. Therefore, the levels of total oocysts reported by Gennaccaro et al. (2003) may be nontarget organisms. One could also argue that the antibodies used to detect intracellular reproductive stages in cell culture may also react with nontarget organisms and not only *C. parvum*. In addition, Rochelle et al. (2001) reported that researchers might be reporting incorrect information based on how they are enumerating infectious foci. The authors reported that in their study, infectious foci were defined as clusters of intracellular developmental stages of *C. parvum*. These foci contained 2–50 developmental stages and had diameters of 12.25–143.5 μm. In other studies, infectious foci were defined as individual developmental stages rather than clusters leading to higher reported numbers of infectious *Cryptosporidium* (Slifko et al., 1999).

The cell culture method, using real-time PCR for detection of infectious *Cryptosporidium* oocysts in seeded IMS concentrates generated from wastewater samples, yielded conflicting information regarding the presence of infectious oocysts. The results seem to suggest that interferences were present or amplification of nontarget DNA lead to the erroneous results. In contrast, the results obtained using FDM show the presence of infectious oocysts in secondary effluent from the plant in California, although seeded oocysts were not infectious. The fluorescent counts for the California wastewater samples show that 34–41 oocysts were detected in 10 L samples of secondary effluent and that 2–6 oocysts were infectious in this volume. A high recovery rate of seeded oocysts detected in sample 3 (139%) may in fact be caused by higher level of indigenous oocysts present in that subsample. Results from the Pennsylvania samples show that <1–1 oocyst were detected in unseeded secondary effluents and no infectious oocysts were detected in these samples.

In conclusion, although much effort has been invested in developing cell culture methods for detection of infectious oocysts in environmental matrices, apparently much more work needs to be focused on overcoming interferences in wastewater. A method that incorporates cell culture is necessary to detect infectious oocysts because only infectious oocysts pose a risk to public health. To do this, development of methods to maximize the removal of interferences from the wastewater matrix is necessary. IMS has shown considerable promise in reducing interferences and also with minimizing the numbers of nontarget organisms that are captured. The reduction of interferences and nontarget organisms improve the chance that cell will not "die" when IMS concentrates are inoculated onto the monolayers. FDM showed that infectious oocysts can be detected in IMS purified secondary effluents. However, a limitation to this method is the sensitivity of the method to infections. Rochelle et al. (2001) reported only a 7% infection rate of HCT-8 cells with oocysts that were fresh and noted decreased rates of infection as the oocysts aged.

These results demonstrate that detecting infectious oocysts in wastewater matrices is difficult and requires more research before this is a standard technique included in wastewater analysis. Research efforts need to be expended on these infectivity and molecular methods for analysis of wastewater effluents because risk assessment cannot be determined otherwise. Using the total oocyst counts detected by microscopic methods is problematic, because likely the majority of the oocysts are not infectious. This is certainly the case in WWTPs employing UV disinfection (Linden et al., 2004).

CHAPTER 7.0

CONCLUSIONS AND IMPLICATIONS FOR WASTEWATER UTILITIES

This research had three major objectives: 1) to develop methods specific for recovery of *Cryptosporidium* oocysts from wastewater matrices, 2) to use the methods in a national survey of WWTPs to develop a database on *Cryptosporidium* occurrence, and 3) to use cell culture and molecular methods to determine if oocysts detected in wastewater matrices were infectious, thus represent a threat to public health.

Reliable methods designed specifically for wastewater samples are available for recovery of *Cryptosporidium* oocysts from the variety of wastewater matrices encountered, from raw sewage to tertiary effluents. The methods have been tested for robustness and, although variability cannot be overcome because of matrix characteristics or chemical interferences not identified in this project, the inclusion of an internal positive control, *Color*Seed, in each sample, permits method performance to be measured routinely. It is imperative that *Color*Seed recovery results not be used to make determinations about the concentrations of indigenous oocysts present in these samples.

Despite considerable effort to develop a method to isolate *Cryptosporidium* oocysts from biosolids, successful trials were accomplished only in biosolids with low percent solids (<10%) and with small amounts of sample (≤5 g). More research effort is needed to develop and optimize *Cryptosporidium* recovery methods with biosolids. *Color*Seed may easily be incorporated into the reported assay as an internal control.

The 15-month occurrence survey showed that *Cryptosporidium* oocysts are commonly found in wastewater samples throughout the treatment process. Issues with sample size, uneven distribution, and matrix characteristics make it difficult to follow the reduction through treatment. As the treatment process improves turbidity and sample quality, making it easier to recover oocysts, it can appear that the process is "adding" rather than removing oocysts. Using analysis of *Cryptosporidium* samples to assess process performance for *Cryptosporidium* removal must be done with these considerations in mind. Looking at the *Cryptosporidium* oocyst removal across the process cannot be done reliably even in drinking water studies unless high seeding levels are used in carefully designed and conducted experiments.

The authors recommend using other parameters that are easily measurable, such as turbidity and solids removal, as indicators of treatment effectiveness. Surrogates such as these are easy to measure, measurements can be continuous to provide real-time data, and subtle changes in the process can be observed as they occur, allowing operators to make the appropriate response immediately. Pathogen occurrence data cannot be used to make operational decisions or protect public health; it can be used to study occurrence over time and develop databases that can be used in planning and design for water treatment.

The two plants that practice tertiary treatment (Vermont and North Carolina) had differing levels of oocysts in their final effluents. For the Vermont plant, oocysts were detected only one occasion in a raw influent sample (11 oocysts/500 mL) in 33 sampling events. In the North Carolina plant, oocysts were detected at low levels throughout the process, and on three of four occasions in the tertiary effluent.

Detecting infectious oocysts in wastewater matrices is difficult and requires more research before this is a standard technique included in wastewater analysis. Oocyst infectivity is critical in risk assessment, because the analytical method measures both live and dead oocysts, and only infectious oocysts present a health risk. Additional research in this area is needed coupled with continued improvement in oocyst recovery methods in wastewater.

APPENDIX A

STANDARD OPERATING PROCEDURE FOR RECOVERING *CRYPTOSPORIDIUM* OOCYSTS FROM RAW SEWAGE AND PRIMARY WASTEWATER EFFLUENTS

A.1 Scope and Application

A.1.1 This method is for the recovery, enumeration, and identification of *Cryptosporidium* oocysts in raw sewage or primary wastewater effluents using concentration, immunomagnetic separation (IMS), and immunofluorescence microscopy. *Cryptosporidium* oocysts are confirmed using 4",6-diamidino-2-phenylindole (DAPI) staining and differential interference contrast (DIC) microscopy.

A.1.2 This method does not identify the species of *Cryptosporidium* or the host species of origin. It cannot determine the viability or infectivity of detected oocysts.

A.2 Summary of Method

A.2.1 *Cryptosporidium* oocysts are recovered from raw sewage or primary wastewater effluents samples by first concentrating the sample after the addition of detergent and *Color*Seed. The sample is concentrated using centrifugation (1500 x *g*, 15 min) and the supernatant discarded. Oocysts are recovered from the sample concentrate utilizing IMS in the presence of kaolin.

A.2.2 The bead-oocyst complex formed during the incubation period is separated from the interfering debris by applying a magnetic field to the sample. The supernatant is discarded. The oocysts are removed from the beads in an acidic medium.

A.2.3 The recovered oocysts are applied to a glass well slide, dried, fixed and then stained well slides with FITC-conjugated mAb and a fluorogenic dye. The stained sample is examined using epifluorescence and DIC microscopy.

A.2.4 Qualitative analysis is performed by examination of each sample for objects that meet the size, shape, and fluorescence characteristics of *Cryptosporidium* oocysts. Potential oocysts are confirmed by their staining characteristics with DAPI and internal morphological characteristics using DIC microscopy. An oocyst is identified when the size, shape, color, and morphology agree with specified criteria.

A.2.5 Quantitative analysis is performed by counting the total number of objects on the slide confirmed as oocysts.

A2.6 Method performance is measured through the addition of *Color*Seed to each sample analyzed. The number of *Color*Seed oocysts recovered serves as an internal control for method performance in these complex matrices because of chemical interferences and the highly variable matrix composition.

A.3 Interferences

A.3.1 Raw sewage and to a lesser extent primary effluents contain high levels of solid and semisolid particles that may inhibit the performance of the IMS beads in a complex matrix. Kaolin is added to bind some of the nonspecific epitopes on these particles allowing the IMS beads to interact with the oocysts present in the samples.

A.3.2 These matrices contain high levels of oil and grease. These fats can coat the surface of the oocysts rendering their surface epitopes inaccessible. Detergent (Tween 80) is added to aid in dissolving grease and oil, thereby exposing the oocyst surface epitope and increasing the potential for capture with the IMS beads.

A.3.3 Chemical interferences (polymers, chlorine, dairy waste, industrial waste, and agricultural runoff) may impact the performance of the IMS beads in recovering oocysts from sewage. The addition of *Color*Seed as an internal control can determine the impact of these interferences on the recovery of oocysts.

A.4 Safety

A.4.1 The biohazard associated with, and the risk of infection from pathogenic bacteria, viruses, and protozoan are high in this method because potentially live and infectious organisms are handled. This method does not purport to address all of the safety problems associated with its use. It is the responsibility of the laboratory to establish appropriate safety and health practices before use of this method. In particular, the analyst/technician must know and observe the safety procedures required in a microbiology laboratory that handles pathogenic organisms while preparing, using, and disposing of sample concentrates, reagents and materials, and while operating sterilization equipment.

A.4.2 Refer to MSDS sheets for information on the correct handling and disposal of chemicals and reagents used in this method.

A.4.3 Laboratory personnel should wear gloves, a labcoat and eye protection when handing these samples and other potentially contaminated equipment and reagents. Gloves must be removed or changed before handling any other laboratory surfaces or equipment.

A.5 Materials and Apparatus

A.5.1 Sample Collection

10 L polypropylene carboy with a wide mouth
2.5 L plastic pail with 15' length of plastic rope
Graduated cylinders, 25 mL, 50 mL, 250 mL, 500 mL, and 1 L

A.5.2 Equipment for sample concentration

- Polystyrene or polypropylene conical tubes with screw caps:250-mL
- Tube racks
- Disposable underpads and gloves
- Wash bottles for buffer solutions
- Large beakers
- Vortexer
- Centrifuge with swinging buckets and adapters for 250 mL conical centrifuge tubes capable of operating at 1500 x g for15 min
- Vacuum pump with aspirating flask and trap
- Pasteur pipettes
- Disposable polystyrene pipettes, 10 mL and 25 mL
- Pipet-Aid

A.5.3 Equipment for IMS and sample staining

- Leighton tubes
- Pasteur pipettes and bulb
- Dynal IMS magnets - MPC-1 and MPC-M
- Sample rotator capable of operating at 15 –25 rpm
- Timer
- Dynal Spot On slides. cat.no. 740.04
- Glass coverslips, 18 mm^2
- Clear nailpolish
- Pipettors, P1000, P100 and P10 with appropriate tips
- Slide warmer capable of operating at 42-50°C
- Dropper bottles for reagents
- Vortexer
- Incubator capable of operating at 34-40°C
- Humid chamber (Rubbermaid container with wet paper toweling)
- Eppendorf tube, 1.5 mL

A.5.4 Equipment for microscopic examination

Epifluorescence/differential interference contrast (DIC) with stage and ocular micrometers and 20X (N.A. = 0.4) to 100X (N.A. = 1.3) objectives—Zeiss Axioskop

Excitation/band-pass filters for immunofluorescent assay (FA)—Zeiss 487909 or equivalent, including, 450- to 490-nm exciter filter, 510 nm dichroic beam-splitting mirror, and D540/40 nm barrier or emission filter

Excitation/band-pass filters for DAPI—filters (Chroma Technology, Brattleboro, VT), or equivalent, including D340/60 nm exciter filter, GG420 nm barrier filter and 400DCLPnm Chromatic Beam Splitter.

Excitation/band pass filters for PI-filters (Chroma Technology, Brattleboro, VT), or equivalent including excitation 546 nm, emission LP590 nm, dichroic FT580 nm)

Nonfluorescing immersion oil
Lens paper without silicone
Lens cleaner

A.6 Reagents and Standards Preparation

A.6.1 Kaolin (Sigma Chemical # K7375).

A.6.2 Dynal Cryptosporidium IMS kit (Dynal # 730.01 or 730-11) containing 10X SL-A buffer and 10X SL-B buffer and anti-*Cryptosporidium* IMS beads.

A.6.3 Phosphate buffered saline (PBS), 0.15 M: Dissolve 1.07 g Na_2HPO_4, 0.39 g NaH_2PO_4◈$2H_2O$, and 8.5 g NaCl in 800 mL DI. Dilute to 1000 mL with DI. Store at room temperature in plastic or glass up to 1 year or until growth is apparent.

A.6.4 0.1 N HCl: Purchased standardized American Chemical Society (ACS) reagent from commercial vendor.

A.6.5 1.0 N NaOH: Purchased standardized ACS reagent from commercial vendor.

A.6.6 Tween 80, 20% (w/v): Dissolve 200 g of Tween 80 in 700 mL of DI. Dilute to 1000 mL when all is dissolved. Store in plastic or glass at room temperature for up to 1 year or until growth is apparent.

A.6.7 Crypt-a-Glo FITC-conjugated mAb from Waterborne, Inc. Prepare and store as prescribed by the manufacturer's instructions.

A.6.8 DABCO mounting medium: Dissolve 2 g DABCO (Sigma Chemical # D2522) in warm solution of 40 mL 150 mM PBS and 60 mL glycerol. Store in glass at room temperature for up to 6 months. Discard after expiration date.

A.6.9 Stock DAPI solution—Add 2.5 mL absolute methanol (Sigma Chemical # 17-5) to 5 mg DAPI (Sigma Chemical D9542). Store at 0–8°C in the dark. No shelf life has been determined for this reagent.

A.6.10 DAPI staining solution (1/5000 dilution in PBS)—Add 10 μL of 2 mg/mL DAPI stock solution to 50 mL of 150 mM PBS. Prepare daily. Store at 0-8°C in the dark except when staining.

A.6.11 *Color*Seed Internal Control Standard. Available from BTF, Ltd.

A.6.12 Tween 80, 0.05%: Dissolve 0.5 mL Tween 80 in 1000 mL DI.

A.7 Calibration

A.7.1 Pipette Calibration: Refer to manufacturer's instructions for calibration frequency.

A.7.2 Microscope set up, calibration, operation and maintenance: Refer to manufacturer's instruction manuals.

A.8 Quality Control

A. 8.1 *Color*Seed Internal Control Standard: Observations made during method development trials showed variability in the wastewater matrix composition from day to day at the same source as well as from subsample to subsample. It was observed that matrix composition could play a role in the performance of the IMS in recovering oocysts from the sample. When dairy waste from the local coop plant is dumped in the waste stream no oocysts are recovered in seeded raw sewage samples. For these reasons the incorporation of *Color*Seed as an internal control for evaluation of method performance was necessary to meet data quality objectives.

A.8.2 New lots of reagents are tested upon receipt to determine if they perform adequately. For example, new IMS kits are tested by seeding 10 mL DI with a predetermined number of *C. parvum* oocysts and processing the sample using the IMS procedure. Recovered oocysts are stained and enumerated using epifluorescence microscopy. The number of oocysts recovered is compared to the number seeded to determine the percent recovery and in effect the efficiency of the IMS kit. Oocysts are stained with new and old lots of FITC-mAb to compare staining characteristics and intensities. If new lots perform as well or better than older approved lots the new lot is acceptable for use in the method.

A.9 Sample Preparation and Procedure

A.9.1 Sample Collection, Preservation, and Hold Times

A.9.1.1 Collect raw sewage samples after the screens if possible so that larger particles and objects are removed. Be sure that the sample collected represents the sewage that is flowing into the treatment plant. If this is not possible composite sampling may be required. Determine the most appropriate sample points and times to obtain the most representative sample. For grab samples, a small pail with a 12' length of plastic rope tied to the handle is lowered into the waste stream. The bucket is allowed to fill and then the pail is raised using the rope. The collected sample is emptied into a 10 L carboy with a wide mouth opening. Repeat the sample collection procedure until an adequate sample volume has been collected. Alternatively, WWTPs with sample taps in the plant may collect grab samples at these points as long as the samples are representative of the wastewater stream.

A.9.1.2 Primary effluent samples are collected from a representative site. The sample collection procedure used is described in 9.1.1.

A.9.1.3 Samples are collected the day of analysis. If same day analysis is not possible, chill sample to 4°C until analysis can begin.

A.9.1.4 If is preferable to have the analysis commence immediately. If this is not possible it should be initiated within 24 h of sample collection.

A.9.2 Sample Analysis for Raw Sewage and Primary Effluents

A.9.2.1 Ensure that the carboy cover is secure and shake the carboy to resuspend settled particles. Decant off sample in to 2 L beaker.

A.9.2.2 Measure a volume of the wastewater sample between 250 mL and 1000 mL into a graduated cylinder and transfer to a 500 or 1000 mL plastic bottle with a screw cap.

A.9.2.3 Add 20% Tween 80 solution to the sewage sample to yield a final concentration of 1%. Therefore, add 12.5 mL to 250 mL sample, 25 mL to a 500 mL sample and 50 mL to a 1000 mL sample.

A.9.2.4 Using the protocol prescribed by the manufacturer of *Color*Seed transfer the contents of 1 vial to the wastewater sample containing the Tween 80 solution.

A.9.2.5 Shake sample gently and decant to clean 250 mL conical centrifuge tubes. After transfer of the entire sample rinse the bottle with copious volumes of PBS to ensure a complete transfer.

A.9.2.6 Balance tubes in centrifuge buckets and cap tightly. Centrifuge the sample at 1500 x g for 15 min. Do not use the brake.

A.9.2.7 Aspirate the supernatant from conical tube to just above the pellet. If more than 1 tube was used for each sample aspirate to a volume in each tube resulting in a total volume not more than 10 mL. For example, if 4 tubes were used for the 1000 mL sample, aspirate each tube to approximately 2.5 mL or to just above the packed pellet. If this is not possible, aspirate to above the packed pellet and determine the total volume of the sample concentrate. Analyze a portion of the sample concentrate containing no more than 4 mL packed pellet volume.

A.9.3 Immunomagnetic Separation (IMS) Procedure

A.9.3.1 Transfer sample concentrate to a Leighton tube containing 1 mL each 10 X SL-A buffer and SL-B buffer.

A.9.3.2 Weigh out 0.75 g Kaolin on to a piece of waxed weighing paper. Carefully fold the weigh paper in the shape of a funnel and gently tap the kaolin into the Leighton tube containing the sample concentrate.

A.9.3.3 Add 100 µL of vortexed *Cryptosporidium* IMS beads to the Leighton tube. Cap with a clean screw cap. Gently mix the kaolin in the tube by rocking the tube back and forth.

A.9.3.4 Place the tube on to the sample rotator and mix at 18 rpm for 1 h at room temperature.

A.9.3.5 Remove the Leighton tube from the sample mixer and place in the MPC-1. Place the tube with the flat side of the Leighton tube facing the magnet. Without removing the tube from the MPC-1, place the magnet side of the MPC-1 downwards, so the tube is horizontal and the flat side of the Leighton tube is facing down. Gently rock the tube by hand end-to-end through approximately 90°, tilting cap-end and base-end of the tube up and down in turn. Continue the tilting action for 1 min with approximately one tilt/s.

A.9.3.6 Allow the sample to stand in the magnet motionless for 3 min. At the end of 3 min, very gently rock the tube in the MPC-1 for another 1 min.

A.9.3.7 Return the MPC-1 to the upright position, tube vertical, with cap at top. Immediately remove cap and decant (pour off) all the supernatant from the tube held in the MPC-1 into a suitable container. Do not shake the tube and do not remove the tube from MPC-1 during this step.

A.9.3.8 Without removing the tube from the MPC-1 add 10 mL PBS down the side of the tube away from the beads. Remove the tube from the magnet and very gently tilt the tube five times to resuspend the beads. Place the tube back in the MPC-1 to repeat steps 9.3.5 through 9.3.7.

A.9.3.9 Remove the tube from the MPC-1 and resuspend the sample in 1-mL 1X SL-buffer-A (prepared from 10X SL-buffer-A stock—supplied). Mix very gently to resuspend all material in the tube. Do not vortex.

A.9.3.10 Using a Pasteur pipette, transfer all the liquid from the sample tube to a labeled, 1.5-mL microcentrifuge tube. Add a fresh 0.5 mL aliquot of 1 X SL-A buffer to the Leighton to remove residual beads possibly present. Add this rinse to the same microfuge tube containing the first rinse. Place the microcentrifuge in the second magnetic particle concentrator (MPC-M), with magnetic strip in place.

A.9.3.11 Without removing the microcentrifuge tube from the MPC-M gently rock/roll the tube through 180° for approximately 1 min with one 180° roll/rock per second. At the end of this step, the beads and oocysts should produce a well-formed brown dot on the back of the tube.

A.9.3.12 Immediately aspirate the supernatant from the tube and cap held in the MPC-M. If more than one sample is being processed, conduct three 180° rock/roll actions before removing the supernatant from each tube. Take care not to disturb the material attached to the wall of the tube adjacent to the magnet. Do not shake the tube. Do not remove tube from MPC-M while conducting these steps.

A.9.3.13 Remove the magnetic strip from the MPC-M. Add 100 µL of 0.1 N HCl, then vortex vigorously for 30 s. Incubate for 10 min at room temperature. Vortex vigorously vial for 10 s. Replace magnetic strip in MPC-M and replace tube in MPC-M. Allow vial to stand for 2 min undisturbed.

A.9.3.14 Remove well slides from the box in the refrigerator and lay the slides on a flat surface for 15 min to allow them to warm to room temperature. Use one slide for each sample being analyzed.

A.9.3.15 Add 10 µL of 1.0 N NaOH to a sample well.

A.9.3.16 Do not remove the microcentrifuge tube from the MPC-M. Transfer the entire sample from the microcentrifuge tube in the MPC-M to the sample well with the NaOH. Do not disturb the beads at the back wall of the tube.

A.9.3.17 Dry the sample on the slide warmer at 42°C. Samples may be air-dried overnight. Proceed with staining the sample as described Section A.9.4.

A.9.3.18 Record all pertinent information for each sample on the laboratory benchshcet.

A.9.4 Sample Staining Procedure

A.9.4.1 Positive and negative controls must be prepared. For the positive control, pipette 200 to 500 intact oocysts onto the center of a well and distribute evenly over the well area. For the negative control, place a drop of 150 mM PBS onto the center of a well and spread it over the well area with a pipette tip.

A.9.4.2 Allow samples and staining controls to evaporate to dryness (30 min to 1 h). Alternately, slides may air-dry overnight.

A.9.4.3 Remove the slides from the slide warmer and apply 1 drop of absolute methanol to each well containing the dried sample and allow the slide to air dry until the methanol has evaporated (3–5 min).

A.9.4.4 Place 1 drop of FITC-mAb to the center of each well, the positive-control well and the negative-control well. If necessary, gently spread FITC-mAb over well using a pipette tip or wooden toothpick. Use a clean tip or toothpick for each sample. Take care to avoid scratching the well surface. Place the slides in a humid chamber and incubate at 37°C for 30–45 min. The humid chamber consists of a tightly sealed plastic container containing damp paper towels on which the slides are placed.

A.9.4.5 After the incubation period, remove the slides from the humid chamber. Use a clean Pasteur pipette tip attached to a vacuum source to gently aspirate the excess FITC-mAb from the side of each well. Use a new pipette for each sample. When performing this step, ensure that the vacuum source is at the absolute minimum (<2 in Hg vac.) and ensure that the pipette tip does not scratch the well surface.

A.9.4.6 Apply a drop of 150 mM PBS, pH 7.2, to each of the 10 wells and allow to stand for 1 min, and then aspirate the excess PBS. When removing the excess PBS, ensure that the pipette tip does not scratch the well surface.

A.9.4.7 Apply a drop of the dilute DAPI solution to the center of each well. Incubate at room temperature for 2 min. Aspirate excess DAPI and repeat 9.4.6.

A.9.4.8 Apply a small drop of mounting medium (2% DABCO in 60% glycerol, 40% 150 mM PBS) to the center of each well.

A.9.4.9 Place coverslip on each well and seal the edges of the coverslip onto the slide using clear nail polish. Store stained slides in the dark at 2–8°C in a humid chamber.

A.9.5 Sample Examination and Enumeration

A.9.5.1 Scanning technique: Scan each well using a systematic side-to-side pattern.

A.9.5.2 Staining Controls

A.9.5.2.1 If the positive control contains oocysts at the appropriate fluorescence for both FA and DAPI, proceed with documenting three oocysts on the positive staining control benchsheet. Record size, DAPI staining characteristics (number of DAPI-stained nuclei, cytoplasmic staining, etc.), and internal morphology using DIC microscopy.

A.9.5.2.2 Scan the negative staining control. Document results on the negative staining control benchsheet. If it does not contain any oocysts, use epifluorescence microscopy to scan the entire well for each sample at no less than 200X total magnification.

A.9.5.3 Sample Examination

A.9.5.3.1 When brilliant apple-green fluorescing ovoid or spherical objects 4–6 μm in diameter or are observed with brightly highlighted edges, switch the microscope to the UV filter block for DAPI, the green filter block for Texas Red, and then to DIC.

A.9.5.3.2 Using the UV filter block for DAPI, the object will exhibit one of the following characteristics:

(a) Up to four distinct, sky-blue nuclei
(b) Intense blue internal staining
(c) Light blue internal staining (no distinct nuclei)
(a) and (b) are recorded as DAPI +; (c) is recorded as DAPI –

A.9.5.3.3 Using DIC, look for external or internal morphological characteristics atypical of *Cryptosporidium* (oo)cysts (e.g., spikes, stalks, appendages, pores, one or two large nuclei filling the cell, red fluorescing chloroplasts, crystals, spores). If atypical structures are not observed, then categorize each apple-green fluorescing object as:

(a) An empty *Cryptosporidium* oocyst
(b) A *Cryptosporidium* oocyst with amorphous structure
(c) A *Cryptosporidium* oocyst with internal structure
(1–4 sporozoites/oocyst)

A.9.5.3.4 Using the green filter set, look for the red rim fluorescence of the *Color*Seed oocysts. If no red fluorescence is observed, the oocyst is indigenous to the sample.

A.9.5.3.5 Record the shape and measurements to the nearest 0.5 μm at 1000–1600X total magnification for each such object. Although not a defining characteristic, surface oocystfolds may be observed in some specimens.

A.9.5.3.6 For each oocyst observed, record the number of sporozoites observed or if amorphous structure was observed. If an oocyst is empty this is also recorded.

A.9.5.3.5 Record the date and time that sample examination and confirmation was completed on the laboratory benchsheet.

A.10 Calculations

A.10.1 To determine the percent recovery of *Color*Seed oocysts divide the number of *Color*Seed oocysts recovered in the sample by the total number seeded into the sample and multiply by 100.

A.11 Acceptance of Data

A.11.1 Not applicable at this time.

A.12 Reporting of Results

A.12.1 Results for indigenous oocysts are reported as the number of oocysts per volume analyzed.

A.12.2 *Color*Seed oocyst recoveries are reported as "% recovery of *Color*Seed oocysts."

A.13 Supplemental Documents

A.13.1 Manual for the maintenance, operation, and calibration of the microscope.

A.13.2 Manual for the operation, maintenance, and calibration of micropipettors.

A.13.3 Instructions and certificate of analysis for *Color*Seed product.

APPENDIX B

STANDARD OPERATING PROCEDURE FOR RECOVERING *CRYPTOSPORIDIUM* OOCYSTS IN SECONDARY AND TERTIARY WASTEWATER EFFLUENTS

B.1 Scope and Application

B.1.1 This method is for the recovery, identification and concentration of *Cryptosporidium* oocysts in secondary and tertiary wastewater effluent samples by filtration, immunomagnetic separation (IMS), and immunofluorescence assay microscopy (FA). *Cryptosporidium* may be confirmed using 4',6-diamidino-2-phenylindole (DAPI) staining and differential interference contrast (DIC) microscopy.

B.1.2 This method will not identify the species of *Cryptosporidium* or the host species of origin, nor can it determine the viability or infectivity of detected oocysts.

B.2 Summary of Method

B.2.1 An Envirochek HV capsule filter is seeded with one vial of *Color*Seed and then 10 L secondary effluent or up to 100 L of tertiary effluent is filtered through the capsule with the oocysts and extraneous materials retained on the filter.

B.2.2 Retained particles are pretreated with a sodium buffer solution followed by a rinse with deionized (DI) water. The oocysts are then removed by elution with an aqueous buffered salt and detergent solution. The salt/detergent solution from the filter is centrifuged to settle the oocysts, and the supernatant fluid is aspirated.

B.2.3 The bead-oocyst complex formed during the incubation period is separated from the interfering debris by applying a magnetic field to the sample. The supernatant is discarded. The oocysts are removed from the beads in an acidic medium.

B.2.4 The recovered oocysts are applied to a glass well slide, dried, fixed and then stained well slides with FITC-conjugated mAb and a fluorogenic dye. The stained sample is examined using epifluorescence and differential interference contrast (DIC) microscopy.

B.2.5 Qualitative analysis is performed by scanning each slide well for objects that meet the size, shape, and fluorescence characteristics of *Cryptosporidium* oocysts. Potential oocysts are confirmed by their staining characteristics with DAPI and DIC microscopy. An oocyst is identified when size, shape, color, and morphology agree with specified criteria and examples in a photographic library.

B.2.6 Quantitative analysis is performed by counting the total number of objects on the slide confirmed as oocysts.

B.2.7 Method performance is measured through the addition of *Color*Seed to each sample. The number of *Color*Seed cysts recovered in each sample serves as an internal control of method performance in these complex matrices because of chemical interferences and the highly variable matrix composition.

B.3 Interferences

B.3.1 Interferences from treatment chemicals may inhibit the ability of the elution buffers to adequately recover oocysts from the filter material. In method development trials, the use of sodium hexametaphosphate as a pretreatment buffer has shown to improve oocyst recoveries in both secondary and tertiary wastewater effluents.

B.3.2 Chemical interferences (polymers, chlorine, dairy waste, industrial waste, agricultural runoff) may affect performance of the IMS beads in recovering oocysts from wastewater. The addition of *Color*Seed as an internal control can determine the impact of interferences to oocyst recovery.

B.4 Safety

B.4.1 The biohazard associated with, and the risk of infection from, oocysts is high in this method because potentially live and infectious organisms are handled. This method does not purport to address all of the safety problems associated with its use. It is the responsibility of the laboratory to establish appropriate safety and health practices before use of this method. In particular, the analyst/technician must know and observe the safety procedures required in a microbiology laboratory that handles pathogenic organisms while preparing, using, and disposing of sample concentrates, reagents and materials, and while operating sterilization equipment.

B.4.2 Refer to MSDS sheets for information on the correct handling and disposal of chemicals and reagents used in this method.

B.4.3 Laboratory personnel should wear gloves, a labcoat and eye protection when handing these samples and other potentially contaminated equipment and reagents. Gloves must be removed or changed before handling any other laboratory surfaces or equipment.

B.5 Apparatus and Materials

B.5.1 Sample filtration

10-L or 50-L graduated carboy with bottom delivery port
Large stir bar
Stir plate
Centrifugal pump
Tubing
Adjustable flow meter – 0.4–4.0 lpm
Pipettors, P1000, P100 and P10 with appropriate tips
Disposable underpads and gloves
Miscellaneous plumbing parts and Teflon tape
Tools: channels locks, sockets drivers, etc.
Envirochek HV, product 12099

B.5.2 Equipment for filter elution and sample concentration

Polystyrene or polypropylene conical tubes with screw caps, 250 mL
Tube racks
Disposable underpads and gloves
Wash bottles for buffer solutions
Large beakers
Wrist-action shaker with arm clamps for agitation of sampling capsules
Tools: socket drivers, channel locks, etc.
Vortexer
Centrifuge capable of operating at 1500 x g and holding 250 mL conical centrifuge tubes
Vacuum pump with side-arm Erlenmeyer flask and moisture trap
Pasteur pipettes
Disposable pipettes. 10 mL
Pipet-Aid

B.5.3 Equipment for IMS and sample staining

Leighton tubes
Pasteur pipettes and bulb

Dynal IMS magnets - MPC-1 and MPC-M
Sample rotator capable of operating at 15 –25 rpm
Timer
Dynal Spot On slides. cat.no. 740.04
Glass coverslips, 18 mm^2
Clear nailpolish
Pipettors, P1000, P100 and P10 with appropriate tips
Slide warmer capable of operating at 42-50°C
Dropper bottles for reagents
Vortexer
Incubator capable of operating at 35-37°C
Humid chamber
Eppendorf tube, 1.5 mL

B.5.4 Equipment for microscopic examination

Epifluorescence/differential interference contrast (DIC) with stage and ocular micrometers and 20X (N.A.= 0.4) to 100X (N.A. = 1.3) objectives— Zeiss Axioskop

Excitation/band-pass filters for immunofluorescent assay (FA)— Zeiss 487909 or equivalent, including, 450- to 490-nm exciter filter, 510 nm dichroic beam-splitting mirror, and D540/40 nm barrier or emission filter

Excitation/band-pass filters for DAPI—filters (Chroma Technology, Brattleboro, VT), or equivalent, including D340/60 nm exciter filter, GG420 nm barrier filter and 400DCLPnm Chromatic Beam Splitter.

Excitation/band pass filters for PI-filters (Chroma Technology, Brattleboro, VT), or equivalent including excitation 546 nm, emission LP590 nm, dichroic FT580 nm)

Nonfluorescing immersion oil

Lens paper without silicone

Lens cleaner

B.6 Reagents and Standards Preparation

B.6.1 Reagent water—water in which oocysts and interfering materials and substances, including magnetic minerals, are not detected by this method.

B.6.2 Laureth-12 Elution Buffer

B.6.2.1 Laureth-12—PPG Industries, Gurnee, IL, cat. no. 06194.

B.6.2.2 1 M Tris, pH 7.4—Dissolve 121.1 g Tris (Fisher cat. no. BP152) in 700 mL of reagent water and adjust pH to 7.4 with 6N HCl or 6N NaOH. Dilute to a final 1000 mL with reagent water and adjust the final pH. Filter-sterilize through a 0.2-µm membrane into a sterile plastic container and store at room temperature for one year.

B.6.2.3 0.5 M EDTA, 2 Na, pH 8.0—Dissolve 186.1 g ethylenediamine tetraacetic acid, disodium salt dihydrate (Fisher cat. no. S311) in 800 mL and adjust pH to 8.0 with 6.0 N HCl or 6.0 N NaOH. Dilute to a final volume of 1000 mL with reagent water and adjust to pH 8.0. Store in a plastic container at room temperature for one year.

B.6.2.4 Antifoam A—Sigma Chemical Co. cat. no. A5633

B.6.2.5 Preparation of buffer solution—Place a beaker containing 1 g Laureth-12 in 100 mL DI on a magnetic hot plate with a stir bar. Heat the solution until the Laureth 12 is melted. Transfer to a 1000-mL graduated cylinder. Rinse the beaker several times to ensure the transfer of the detergent to the cylinder. Add 10 mL of Tris solution, pH 7.4; 2 mL of EDTA solution, pH 8.0; and 150 µL Antifoam A. Dilute to 1000 mL with reagent water. Store at room temperature and discard after 1 month.

B.6.3 Dynal Cryptosporidium IMS kit (Dynal # 730.01 or 730-11) containing 10X SL-A buffer and 10X SL-B buffer and anti-*Cryptosporidium* IMS beads.

B.6.4 Phosphate buffered saline (PBS), 0.15 M: Dissolve 1.07 g Na_2HPO_4, 0.39 g NaH_2PO_4◈ $2H_2O$, and 8.5 g NaCl in 800 mL DI. Dilute to 1000 mL with DI. Store at room temperature in plastic or glass up to 1 year or until growth is apparent.

B.6.5 0.1 N HCl: Purchased standardized ACS reagent from commercial vendor.

B.6.6 1.0 N NaOH: Purchased standardized ACS reagent from commercial vendor.

B.6.7 Tween 80, 20% (w/v): Weigh 200 g of Tween 80 in a 2 L glass beaker. Add 700 mL of DI. Place the beaker on a magnetic stir plate and place a large stir bar in the beaker. Dissolve the Tween with stirring. When the Tween has dissolved dilute to 1000 mL. Store in plastic or glass at room temperature for up to 1 year or until growth is apparent.

B.6.8 Crypt-a-Glo FITC-conjugated mAb from Waterborne, Inc. (Catalog # 400FL). Prepare and store as prescribed in the manufacturer's instructions.

B.6.9 DABCO mounting medium: Dissolve 2 g DABCO (Sigma Chemical # D2522 in warm solution of 40 mL 150 mM PBS and 60 mL glycerol. Store in glass at room temperature for up to six months. Discard after expiration date.

B.6.10 Stock DAPI solution—Add 2.5 mL absolute methanol (Sigma Chemical # 17-5) to 5 mg DAPI (Sigma Chemical D9542). Store at 0–8°C in the dark. No shelf life has been determined for this reagent.

B.6.11 DAPI staining solution (1/5000 dilution in PBS)—Add 10 µL of 2 mg/mL DAPI stock solution to 50 mL of 150 mM PBS. Prepare daily. Store at 0–8°C in the dark except when staining.

B.6.12 *Color*Seed Internal Control Standard. Available from BTF, Ltd.

B.6.13 Tween 80, 0.05%: Dissolve 0.5 mL Tween 80 in 1000 mL DI.

B.6.14 Sodium hexametaphosphate solution (NaHMP), 5%: dissolve 50 g $(NaPO_4)_x \cdot Na_2O$ (where $x \geq 13$) (Fisher Scientific # S333 or equivalent) in 800 mL DI. Dilute to 1000 mL. Store in glass or plastic at room temperature for up to one year. Discard when expiration date is reached or if growth is apparent.

B.7 Calibration

B.7.1 Pipette Calibration: Refer to manufacturer's instructions for calibration frequency.

B.7.2 Microscope set up, calibration, operation and maintenance: Refer to manufacturer's instruction manuals.

B.8 Quality Control

B.8.1 Initial Precision and Recovery IPR

B.8.2 *Color*Seed Internal Control Standard: Observations made during method development trials showed variability in the wastewater matrix composition from day to day at the same source as well as from subsample to subsample. It was observed that matrix composition could play a role in the performance of the IMS in recovering oocysts from the sample. When dairy waste from the local coop plant is dumped in the waste stream no oocysts are recovered in seeded raw sewage samples. For these reasons the incorporation of *Color*Seed as an internal control for evaluation of method performance was necessary to meet data quality objectives.

B.8.3 New lots of reagents are tested upon receipt to determine if they perform adequately. For example, new IMS kits are tested by seeding 10 mL DI with a predetermined number of *C. parvum* oocysts and processing the sample using the IMS procedure. Recovered oocysts are stained and enumerated using epifluorescence microscopy. The number of oocysts recovered is compared to the number seeded to determine the percent recovery and in effect the efficiency of the IMS kit. Oocysts are stained with new and old lots of FITC-mAb to compare staining characteristics and intensities. If new lots perform as well or better than older approved lots the new lot is acceptable for use in the method.

B.9 Sample Preparation and Procedure

B.9.1 Carboy spiking in the laboratory.

B.9.1.1 Arrange a bottom-dispensing carboy (10 or 50 L)to gravity-feed an Envirochek™ HV capsule filter with the outlet attached to a self-priming centrifugal pump.

B.9.1.2 Place a large stir bar in the carboy. Fill the carboy with 10 L of secondary effluent or 50 L of tertiary effluent. Place the carboy on the stir plate. Turn the stirrer on so that the bar creates a vortex.

B.9.1.3 Add 1 vial of *Color*Seed as described in the manufacturer's protocol.

B.9.1.4 Install filter by securing the inlet and outlet ends with the appropriate clamps/fittings.

B.9.1.5 Turn on the pump and allow the flow rate to stabilize. Set flow at $2\ L \cdot min^{-1}$. As the carboy is depleted, check the flow rate and adjust if necessary.

B.9.1.6 When the water level approaches the discharge port of the carboy, turn off the stirrer and tilt the carboy so that it is completely emptied. At that time, turn off the pump and add 1 L of PBST to the 10 L carboy or 5 L to a 50 L carboy. Swirl the contents to rinse down the sides.

B.9.1.7 Turn on the pump. Allow the pump to pull all the water through the filter and turn off the pump.

B.9.1.8 Disconnect the inlet end of the capsule filter assembly while maintaining the level of the inlet fitting above the level of the outlet fitting to prevent backwashing and the loss of oocysts from the filter. Restart the pump and allow as much water to drain as possible. Turn off the pump.

B.9.1.9 Cap the inlet and outlet ports of the filter. Refrigerate at 2–8°C until ready to process.

B.9.1.10 Samples must be processed within 24 h of filtration.

B.9.2 Sample collection in the field.

B.9.2.1 Add 1 vial of *Color*Seed to an Envirochek HV capsule filter containing reagent water using the protocol prescribed by the manufacturer through the inlet port. Cap and store filter at 2–8°C for transport to WWTP.

B.9.2.2 Sample the tertiary effluent at a point that is representative of the effluent leaving the plant. Filter tertiary effluent through the capsule filter at a flow rate of $2\ L \cdot min^{-1}$ until a volume up to 100 L has been collected on the filter or until the flow rate is less than $0.5\ L \cdot min^{-1}$.

B.9.2.3 Cap the inlet and outlet ports and transport to the laboratory and refrigerate at 2–8°C until ready to process samples.

B.9.2.4 Samples must be processed within 24 h of filtration.

B.9.3 Elution Procedure

B.9.3.1 Assemble the wrist-action shaker with the clamps aligned vertically so that the filters will be aligned horizontally.

B.9.3.2 Designate at least one 250-mL conical centrifuge tube for each sample and label with the sample number.

B.9.33 Record all pertinent sampling information on the laboratory bench sheet. Also record the name of the analyst performing the elution on the bench sheet.

B.9.3.4 Pour 5% NaHMP buffer through the inlet fitting. Add enough buffer to cover the pleated membrane. Replace the inlet cap and clamp the cap in place.

B.9.3.5 Securely clamp the capsule in one of the clamps on the wrist-action shaker with the bleed valve positioned at the top on a vertical axis (in the 12 o'clock position). Turn on the shaker and set the speed to maximum speed for approximately 5 min.

B.9.3.6 Remove the capsule from the wrist shaker and remove the solution from the capsule by attaching the outlet to a vacuum source with a catch flask. Replace the outlet cap when the solution has been completely removed from the filter.

B.9.3.7 Fill the capsule with DI through the inlet port. Immediately remove the DI through the outlet by attaching the outlet to a vacuum source with a catch flask.

B.9.3.8 Fill the rinsed capsule filter with L-12 buffer through the inlet port to above the pleated membrane.

B.9.3.9 Place the filter on the wrist shaker with the bleed valve in the 12 o'clock position. Turn the shaker on at maximum speed for 5 min.

B.9.3.10 Remove the filter from the shaker, remove the inlet cap, and pour the contents of the capsule into a 250-mL conical centrifuge tube.

B.9.3.11 Add another volume of L-12 buffer to cover the pleated membrane. Return the capsule to its clamp on the shaker with the bleed valve positioned at 4 o'clock. Turn on the shaker and agitate the capsule for approximately 5 min.

B.9.3.12 Without decanting the elution buffer rotate the capsule to the 8'oclock position and agitate the filter for another 5 min. Decant contents of capsule to a new centrifuge tube.

B.9.4 Sample concentration

B.9.41 Centrifuge the filter eluate at 1500 x g for 15 min.

B.9.42 Aspirate the supernatant to approximately 10 mL.

B.9.43 Record the packed pellet volume.

B.9.5 Immunomagnetic separation (IMS) procedure.

B.9.5.1 Prepare a 1X dilution of SL-buffer-A from the 10 X SL-buffer-A (colorless solution) supplied. Use DI water as the diluent. For every 1 mL of 1X SL-buffer-A required, take 100 μL of 10 X SL-buffer-A and make up to 1 mL with the DI water. 1 mL of 1X SL-buffer-A will be required per sample or subsample on which the Dynal IMS procedure is performed.

B.9.5.2 To a Leighton tube, add 1 mL of the 10 X SL-buffer-A and 1 mL of the 10 X SL-buffer- B. Transfer the water sample concentrate to the Leighton tube containing the SL-buffers. Rinse the conical tube with a volume of L12 buffer so that the total volume of sample concentrate and rinses does not exceed 10 mL. Label the tube with the sample number.

B.9.5.3 Cap the tube and vortex the Dynabeads® anti-*Cryptosporidium* for approximately 10 s to suspend the beads. Ensure that the beads are fully resuspended by inverting the tube and making sure that there is no residual pellet at the bottom. Add 100 μL of the resuspended beads to the sample tube containing the water sample concentrate and SL-buffer.

B.9.5.4 Affix the sample tube to a rotating mixer and rotate at 18-25 rpm for 1 hour.

B.9.5.5 After rotating for 1 h remove the tube from mixer and place the tube with the flat side of the Leighton tube facing the magnet. Without removing the tube from the MPC-1, place the magnet side of the MPC-1 downwards, so the tube is horizontal and the flat side of the Leighton tube is facing down. Gently rock the tube by hand end-to-end through approximately 90°, tilting cap-end and base-end of the tube up and down in turn. Continue the tilting action for 2 min with approximately one tilt/s.

B.9.5.6 Ensure that the tilting action is continued throughout this period to prevent binding of low-mass materials that are magnetic or magnetizable. If the sample in the MPC-1 is allowed to stand motionless for more than 10 s, repeat step 9.5.5 before continuing procedure.

B.9.5.7 Return the MPC-1 to the upright position, tube vertical, with cap at top. Immediately remove cap and decant (pour off) all the supernatant from the tube held in the MPC-1 into a suitable container. Do not shake the tube and do not remove the tube from MPC-1 during this step.

B.9.5.8 Without removing the tube from the MPC-1 add 10 mL PBS down the side of the tube away from the beads. Remove the tube from the magnet and very gently tilt the tube five times to resuspend the beads. Place the tube back in the MPC-1 to repeat steps 9.5.5 through 9.5.7.

B.9.5.9 Remove the tube from the MPC-1 and resuspend the sample in 1-mL 1X SL-buffer-A (prepared from 10X SL-buffer-A stock—supplied). Mix very gently to resuspend all material in the tube. Do not vortex.

B.9.5.10 Quantitatively transfer all the liquid from the sample tube to a labeled, 1.5-mL microcentrifuge tube. Add a fresh 0.5 mL aliquot of 1 X SL-A buffer to the Leighton to remove residual beads possibly present. Add this rinse to the same microfuge tube containing the first rinse. Place the microcentrifuge in the second magnetic particle concentrator (MPC-M), with magnetic strip in place.

B.9.5.11 Without removing the microcentrifuge tube from the MPC-M gently rock/roll the tube through 180° for approximately 1 min with one 180° roll/rock per second. At the end of this step, the beads and oocysts should produce a well-formed brown dot on the back of the tube.

B.9.5.12 Immediately aspirate the supernatant from the tube and cap held in the MPC-M. If more than one sample is being processed, conduct three 180° rock/roll actions before removing the supernatant from each tube. Take care not to disturb the material attached to the wall of the tube adjacent to the magnet. Do not shake the tube. Do not remove tube from MPC-M while conducting these steps.

B.9.5.13 Remove the magnetic strip from the MPC-M. Add 100 μL of 0.1 N HCl, then vortex vigorously for 30 s. Incubate for 10 min at room temperature. Vortex vigorously vial for 15 s. Replace magnetic strip in MPC-M and replace tube in MPC-M. Allow vial to stand for 2 min undisturbed.

B.9.5.14 Remove 12-mm well slides from the box in the refrigerator and lay the slides on a flat surface for 15 min to allow them to warm to room temperature.

B.9.5.15 Add 10 μL of 1.0 N NaOH to a sample well.

B.9.5.16 Do not remove the microcentrifuge tube from the MPC-M. Transfer the entire sample from the microcentrifuge tube in the MPC-M to the sample well with the NaOH. Do not disturb the beads at the back wall of the tube.

B.9.5.17 Dry the sample on the slide warmer at 42°C. Samples may be air-dried overnight. Proceed with staining the sample as described Section 9.6.

B.9.5.18 Record all pertinent information on the laboratory bench sheet.

B.9.6 Sample Staining

B.9.6.1 Positive and negative controls must be prepared. For the positive control, pipette 200 to 500 intact oocysts onto the center of a well and distribute evenly over the well area. For the negative control, pipette 75 μL of 150 mM PBS onto the center of a well and spread it over the well area with a pipette tip.

B.9.6.2 Place the well slides containing the samples and staining controls on a 42°C slide warmer and evaporate to dryness (approximately 30 min to 1 h). Slides may air-dry overnight.

B.9.6.3 Remove the slides from the slide warmer and apply 1 drop of absolute methanol to each well containing the dried sample and allow the slide to air dry until the methanol has evaporated (3–5 min).

B.9.6.4 Overlay the 10 spike-dose wells, the positive-control well and the negative-control well with FITC-mAb. Place the slides in a humid chamber and incubate at 37°C for 30–45 min. The humid chamber consists of a tightly sealed plastic container containing damp paper towels on which the slides are placed.

B.9.6.5 After the incubation period, remove the slides from the humid chamber. Use a clean Pasteur pipette tip attached to a vacuum source to gently aspirate the excess FITC-mAb from the side of each well. Use a new pipette for each sample. When performing this step, ensure that the vacuum source is at the absolute minimum (<2 in Hg vac.) and ensure that the pipette tip does not scratch the well surface.

B.9.6.6 Apply a drop of 150 mM PBS, pH 7.2, to each well and allow standing for 1 min, and then aspirating the excess PBS. When removing the excess

PBS, ensure that the pipette tip does not scratch the well surface.

B.9.6.7 Apply a drop of the dilute DAPI solution and incubate at room temperature for 2 min. Aspirate excess DAPI and repeat step 9.6.6.

B.9.6.8 Apply a small drop of mounting medium (2% DABCO in 60% glycerol, 40% 150 mM PBS) to the center of each well.

B.9.6.9 Place an 18-mm^2 coverslip on each well containing a sample. Seal the edges of the coverslip onto the slide by using clear nail polish. Store stained slides in the dark at room temperature for short-term storage (5–7 days) or at 2--8°C for long-term storage. Record enumeration results on the benchsheet.

B.9.7 Sample Examination and Enumeration

B.9.7.1 Scanning technique: Scan each well using a systematic side-to-side pattern.

B.9.7.2 SExamination and confirmation using immunofluorescence assay (FA), 4',6-diamidino-2-phenylindole (DAPI) staining characteristics, and differential interference contrast (DIC) microscopy. Record examination and confirmation results on the laboratory benchsheet.

B.9.8 Staining Controls

B.9.8.1 If the positive control contains oocysts at the appropriate fluorescence for both FA and DAPI, proceed with documenting three oocysts on the positive staining control benchsheet.

B.9.8.2 Scan the negative staining control. Document results on the negative staining control benchsheet. If it does not contain any oocysts, use epifluorescence microscopy to scan the entire well for each sample at no less than 200X total magnification.

B.9.9 Sample Examination

B.9.9.1 When brilliant apple-green fluorescing ovoid or spherical objects 4–6 μm in diameter or are observed with brightly highlighted edges, switch the microscope to the UV filter block for DAPI, then to DIC.

B.9.9.2 Using the UV filter block for DAPI, the object will exhibit one of the following characteristics:

(a) Up to four distinct, sky-blue nuclei
(b) Intense blue internal staining
(c) Light blue internal staining (no distinct nuclei)
(a) and (b) are recorded as DAPI +; (c) is recorded as DAPI –

B.9.9.3 Using DIC, look for external or internal morphological characteristics atypical of *Cryptosporidium* (oo)cysts (e.g., spikes, stalks, appendages, pores, one or two large nuclei filling the cell, red fluorescing chloroplasts, crystals, spores). If atypical structures are not observed, then categorize each apple-green fluorescing object as:
(a) An empty *Cryptosporidium* oocyst
(b) A *Cryptosporidium* oocyst with amorphous structure
(c) A *Cryptosporidium* oocyst with internal structure (1–4 sporozoites/ oocyst)

B.9.9.4 Using the Texas Red filter, look for the red rim fluorescence of the *Color*Seed oocysts. If no red fluorescence is observed, the oocyst is indigenous.

B.9.9.5 Record the shape and measurements to the nearest 0.5 μm at 1000–1600X total magnification for each such object. Although not a defining characteristic, surface oocyst folds may be observed in some specimens.

B.9.9.6 For each oocyst observed, record the number of sporozoites observed or if amorphous structure was observed. If an oocyst is empty this is also recorded.

B.9.9.7 Record the date and time that sample examination and confirmation was completed on the laboratory benchsheet.

B.10 Calculations

B.10.1 To determine the percent recovery of *Color*Seed oocysts divide the number of *Color*Seed oocysts recovered in the sample by the total number seeded into the sample and multiply by 100.

B.11 Acceptance of Data

B.11.1 Not applicable at this time.

B.12 Reporting of Results

B.12.1 Results for naturally occurring oocysts are reported as the number of oocysts per volume analyzed.

B.12.2 *Color*Seed oocyst recoveries are reported as "% recovery of *Color*Seed oocysts".

B.13 Supplemental Documents

B.13.1 Manual for the maintenance, operation, and calibration of the microscope.

B.13.2 Manual for the operation, maintenance, and calibration of micropipettors.

B.13.3 Instructions and certificate of analysis for *Color*Seed product.

APPENDIX C

LOG REMOVAL TABLES FOR WWTPS

Table C-1. Log Removal of *Cryptosporidium* Oocysts at Alabama Plant #1.

Month/Year	Concentration of oocysts/L in Raw Influent	Log 10	Concentration of Oocysts/L in Primary Effluent	Log 10	Concentration of oocysts /L in final effluent	Log10	Log Removal Raw/Final	Log Removal Primary/ Final
May 2002	<2	0.30	2	0.30	<0.1	-1	-	>1.3
June 2002	4	0.60	<2	0.30	<0.1	-1	>1.6	-
July 2002	<2	0.30	8	0.90	<0.1	-1	-	>1.9
August 2002	4	0.6	2	0.30	<0.1	-1	>1.6	>1.3
September 2002	<2	0.30	8	0.90	0.3	-0.52	<0.8	1.4
October 2002	24	1.38	2	0.30	<0.1	-1	>2.4	>1.3
November 2002	2	0.30	<2	0.30	0.2	-0.70	1.0	<1.0
December 2002	2	0.30	6	0.78	1.0	0	0.3	0.8
January 2003	<2	0.30	<2	0.30	0.5	-0.30	<0.6	<0.6
March 2003	<2	0.30	2	0.30	0.7	-0.15	<0.5	0.5
April 2003	8	0.90	2	0.30	0.3	-0.52	1.4	0.8

Table C-2. Log Removal of *Cryptosporidium* Oocysts at Alabama Plant #2.

Month/Year	Concentration of oocysts/L in Raw Influent	Log 10	Concentration of Oocysts/L in Primary Effluent	Log 10	Concentration of oocysts /L in final effluent	Log10	Log Removal Raw/Final	Log Removal Primary/ Final
May 2002	<2	0.30	<2	0.30	<0.1	-1.0	-	-
June 2002	<2	0.30	82	1.91	14.5	1.16	Inc	0.8
July 2002	<2	0.30	10	1.0	<0.1	-1.0	-	>2.0
August 2002	<2	0.30	30	1.48	0.3	-0.52	<0.8	2.0
September 2002	<2	0.30	<2	0.30	0.6	-0.22	<0.5	<0.5
October 2002	2	0.30	<2	0.30	<0.1	-1.0	>1.3	-
November 2002	<2	0.30	<2	0.30	0.6	-0.22	<0.5	<0.5
December 2002	2	0.30	2	0.30	9	0.95	Inc	Inc
January 2003	<2	0.30	<2	0.30	2.7	0.43	Inc	Inc
March 2003	<2	0.30	<2	030	<0.1	-1.0	-	-
April 2003	<2	0.30	2	0.30	0.2	-0.70	<1.0	1.0

Table C-3. Log Removal of *Cryptosporidium* Oocysts at Alabama Plant #3.

Month/Year	Concentration of oocysts/L in Raw Influent	Log 10	Concentration of Oocysts/L in Primary Effluent	Log 10	Concentration of oocysts /L in final effluent	Log10	Log Removal Raw/Final	Log Removal Primary/ Final
May 2002	<2	0.30	2	0.30	1.5	0.18	<0.1	0.1
June 2002	2	0.30	<2	0.30	0.3	-0.52	0.8	<0.8
July 2002	<2	0.30	<2	0.30	<0.1	-1.0	-	-
August 2002	<2	0.30	<2	0.30	0.3	-0.52	<0.8	<0.8
September 2002	2	0.30	<2	0.30	0.2	-0.70	1.0	<1.0
October 2002	<2	0.30	<2	0.30	1.2	0.08	<0.2	<0.2
November 2002	2	0.30	2	0.30	0.2	-0.70	1.0	1.0
December 2002	<2	0.30	<2	0.30	3.6	0.56	Inc	Inc
January 2003	<2	0.30	<2	0.30	<0.2	-0.70	-	-
March 2003	<2	0.30	<2	0.30	0.5	-0.30	<0.6	<0.6
April 2003	<2	0.30	<2	0.30	<0.1	-1.0	-	-

Table C-4. Log Removals of *Cryptosporidium* Oocysts at California Plant.

Month/Year	Concentration of Oocysts/L in Raw Influent	Log 10	Concentration of Oocysts/L in Primary Effluent	Log10	Concentration of Oocysts/L in Final Effluent	Log 10	Log Removal Raw/Final	Log Removal Primary /Final
October 2002	24	1.38	44	1.64	<0.1	-1.0	>2.4	>2.6
November 2002	2	0.30	12	1.08	<0.1	-1.0	>1.3	>2.1
December 2002	10	1.00	86	1.93	38.9	1.59	Inc	0.3
January 2003	<2	0.30	14	1.15	0.4	-0.4	<0.7	1.5
February 2003	<2	0.30	2	0.30	2.1	0.32	Inc	Inc
March 2003	<2	0.30	52	1.72	5.7	0.76	Inc	1.0
April 2003	6	0..78	8	0.90	17.3	1.24	Inc	Inc
May 2003	<2	0.30	2	0.30	1.9	0.28	<0	0.0
June 2003	2	0.30	12	1.08	4.4	0.64	inc	0.4

Table C-5. Log Removal of *Cryptosporidium* Oocysts at Colorado Plant.

Month/Year	Concentration of Oocysts/L in Raw Influent	Log 10	Concentration of Oocysts/L in Primary Effluent	Log10	Concentration of Oocysts/L in Final Effluent	Log 10	Log Removal Raw/Final	Log Removal Primary/ Final
October 2002	<2	0.30	<2	0.30	0.23	-0.64	>0.94	>0.94
November 2002	<2	0.30	<2	0.30	0.5	-0.30	>0.60	>0.60
December 2002	<2	0.30	2	0.30	0.21	-0.68	>0.98	>0.78
January 2003	<2	0.30	<2	0.30	0.2	-0.70	>1.00	>1.00
February 2003	<2	0.30	<2	0.30	0.1	-1.00	>1.30	>1.30
March 2003	<2	0.30	2	0.30	0.1	-1.00	>1.30	1.30
April 2003	<2	0.30	24	1.38	0.2	-0.70	>1.00	2.08
May 2003	<2	0.30	<2	0.30	0.7	-0.15	>0.46	>0.46

Table C-6. Log Removals of *Cryptosporidium* Oocysts at North Carolina Plant.

Month/Year March 2003	Concentration of Oocysts/L in Raw Influent	Log 10	Concentration of Oocysts/L in Secondary Effluent	Log 10	Concentration of Oocysts/L in Final Effluent	Log 10	Log Removal Raw/Final	Log Removal Secondary/ Final
March 2003	2	0.30	<0.08	-1.10	0.08	-1.10	1.4	0
April 2003	<2	0.30	0.18	-0.74	0.26	-0.59	<0.9	Inc
May 2003	<2	0.30	0.04	-1.40	<0.10	-1.00	-	-
June 2003	<2	0.30	0.62	-0.21	0.13	-0.89	<1.2	0.7

Table C-7. Log Removal of *Cryptosporidium* Oocyst in Pennsylvania Plant #1.

Month/Year	**Concentration of Oocysts/L in Raw Influent**	Log 10	**Concentration of Oocysts/L in Primary Effluent**	Log 10	**Concentration of Oocysts/L in Final Effluent**	Log 10	Log Removal Raw/Final	Log Removal Primary/ Final
November 2002	**<2**	**0.30**	**<2**	**0.30**	**1.2**	**0.08**	**<0.2**	**<0.2**
December 2002	**<2**	**0.30**	**4**	**0.60**	**3.2**	**0.50**	**Inc**	**0.1**
January 2003	**6**	**0.78**	**2**	**0.30**	**<0.1**	**-1.00**	**>1. 8**	**>1.3**
April 2003	**2**	**0.30**	**<2**	**0.30**	**1.3**	**0.11**	**0.2**	**<0.2**
May 2003	**<2**	**0.30**	**8**	**0.90**	**0.8**	**-0.1**	**<0.4**	**1.0**
June 2003	**<2**	**0.30**	**4**	**0.60**	**2.2**	**0.34**	**inc**	**0.3**

Table C-8. Log Removal of *Cryptosporidium* Oocysts in Pennsylvania Plant #2.

Month/Year	Concentration of Oocysts/L in Raw Influent	Log 10	Concentration of Oocysts/L in Primary Effluent	Log 10	Concentration of Oocysts/L in Final Effluent	Log 10	Log Removal Raw/Final	Log Removal Primary/ Final
November 2002	<2	0.30	No data	-	<0.1	-1.00	-	-
December 2002	<2	0.30	2	0.30	0.70	-0.15	<0.5	0.5
January 2003	4	0.60	<2	0.30	0.20	-0.70	1.3	<1.0
March 2003	<2	0.30	<2	0.30	<0.1	-1.0	-	-
April 2003	<2	0.30	<2	0.30	<0.15	-0.82	-	-
May 2003	<2	0.30	<2	0.30	0.30	-0.52	<0.8	<0.8
June 2003	<2	030	<2	0.30	<0.10	-1.00	-	-

Table C-9. Log Removal of *Cryptosporidium* Oocysts in Pennsylvania Plant #3.

Month/ Year	Concentration of Oocysts/L in Raw Influent	Log 10	Concentration of Oocysts/L in Primary Effluent	Log 10	Concentration of Oocysts/L in Final Effluent	Log 10	Log Removal Raw/Final	Log Removal Primary/ Final
November 2002	2	0.30	<2	-	<0.1	-1.0	>1.3	-
December 2002	<2	0.30	<2	0.30	0.24	-0.62	<0.9	<0.9
January 2003	6	0.78	6	0.30	3.7	0.57	0.2	0.2
March 2003	<2	0.30	4	0.30	<0.2	0.30	-	>1.3
April 2003	2	0.30	<2	0.30	<0.2	0.30	>1.0	-
May 2003	6	0.78	2	0.30	1.0	0.0	0.8	0.3
June 2003	2	0.30	8	0.30	<2.9	0.46	-	>0.4

REFERENCES

Allen, M.J.; Clancy, J.L.; Rice, E.W. The Plain, Hard Truth about Pathogen Monitoring. *J. Am. Water Works Assoc.* **2000**. *92*(9), 64–76.

Anonymous. *Methods for the Examination of Waters and Associated Materials. Isolation and Identification of* Giardia *Cysts,* Cryptosporidium *Oocysts and Free Living Pathogenic Amoebae in Water, etc. 1989;* Dept. of Environment, Standing Committee of Analysts; H.M.S.O. Publication: London, 1990.

Arrowood, M.J.; Donaldson, K. Improved Purification Methods for Calf Derived *Cryptosporidium parvum* oocysts Using Discontinuous Sucrose and Cesium Chloride Gradients. *J. Eukaryot. Microbiol.* **1996**, *43*, S89.

Arrowood, M.J.; Xie; L.; Hurd, M.R. *n Vitro* Assays of Maduramicin Activity against *Cryptosporidium parvum*. *J. Euk. Microbiol.* **1994**, *41*, 23S.

Atherholt, T.B.; Korn, L. ICR Protocol: Alternative Treatment of Parasite Sample Data. *J. Am. Water Works Assoc.* **1999**, *91* (3), 95–102.

Bukhari, Z. CEC, Inc., St. Albans, Vermont. Unpublished observations, 1997.

Bukhari, Z.; Smith, H.V. Effect of Three Concentration Techniques on Viability of *Cryptosporidium parvum* Oocysts Recovered from Bovine Faeces. *J. Clin. Microbiol.* **1995**, *33* (10), 2592–2595.

Bukhari, Z.; Smith, H.V.; Sykes, N.; Humphreys, S.W.; Paton, C.A.; Girdwwod, R.W.A.; Fricker, C.R. Occurrence of *Cryptosporidium* sp. Oocysts and *Giardia* sp. Cysts in Sewage Influents and Sewage Effluents from Sewage Treatment Plants in England. *Water Sci. Technol.* **1997**, *35* (11–12), 385–390.

Bukhari, Z.; McCuin, R.M.; Fricker, C.R.; Clancy, J.L. Immunomagnetic Separation of *Cryptosporidium parvum* from Source Waters Samples of Various Turbidities. *Appl. Environ. Microbiol.* **1998,** *64* (5), 4495–4499.

Carraro, E.; Fea, E.; Salva, S.; Gilli. G. Impact of Wastewater Treatment on *Cryptosporidium* Oocysts and *Giardia* Cysts Occurring in a Surface Water. *Water Sci. Technol.* **2000**, *41* (7), 31–37.

Champliaud, D.; Gobet, P.; Naciri, M.; Vagner, O.; Lopez, J.; Buisson, J.; Varga, I.; Harly, G.; Mancassola, R.; Bonnin, A. Failure to Differentiate *Cryptosporidium parvum* from *C.*

meleagridis Based on PCR Amplification of Eight DNA Sequences. *Appl. Environ. Microbiol.* **1998**, *64* (4), 1454–1458.

Chauret, C.; Springthorpe, S.; Sattar, S. Fate of *Cryptosporidium* Oocysts, *Giardia* Cysts and Microbial Indicators during Wastewater Treatment and Anaerobic Digestion. *Can. J. Microbiol.* **1999**, *45* (3), 257–262.

Clancy, J.L.; Gollnitz, W.D.; Tabib, Z. Commercial Labs: How Accurate Are They? *J. Am. Water Works Assoc.* **1994**, *86* (5), 86–89.

Clancy, J.L.; Hargy, T.; Schaub, S. Proceedings of the International Symposium on *Cryptosporidium* and Cryptosporidiosis. In *Proceedings 1997 International Symposium on Waterborne Cryptosporidium,* Fricker, C.R.; Clancy, J.L.; Rochelle, P., Eds.; American Water Works Association: Denver, Colo., 1997; pp. 87–90.

Connell, K.; Rodgers, C.C.; Shank-Givens, H.L.; Scheller, J.; Pope, M.L.; Miller, K. Building a better protozoan data set. *J. Am. Water Works Assoc.* **2000**, *92* (10), 30–43.

D'Antonio, R.G.; Winn, R.E.; Taylor, J.P.; Gustafson, T.L.; Current, W.L.; Rhodes, G.W. Gary, J.L.; Zajac, R.A. 1985. A Waterborne Outbreak of Cryptosporidiosis in Normal Hosts. *Ann. Intern. Med.* **1985**, *103* (6, Pt. 1), 886–888.

Di Giovanni, G.D.; Hashemi, F.H.; Shaw, N.J.; Abrams, F.A.; LeChevallier, M.W.; Abbaszadegan, M. Detection of Infectious *Cryptosporidium parvum* Oocysts in Surface and Filter Backwash Water Samples by Immunomagnetic Separation and Integrated Cell Culture-PCR. *Appl. Environ. Microbiol.* **1999**, *65* (8), 3427–3432.

Di Giovanni, G.D.; Karim, M.R. LeChevallier, M.W.; Boutros, S.N.; Chandler, J.S.; Abrams, F.A.; Spinner; M.L.; Weihe, J.R. *Overcoming Molecular Sample Processing Limitations: Quantitative PCR;* Project 00-HHE-2b; Water Environment Research Foundation and International Water Association: Alexandria, VA, 2002.

Dumoutier, N.; Mandra, V. *Giardia* and *Cryptosporidium* Removal by Water Treatment Plants. *Water Supply* **1996**. *14* (3/4), 387–401.

DuPont, H.L.; Chappell, C.L.; Sterling, C.R.; Okhuysen, P.C.; Rose, J.B.; Jakubowski, W. 1995. The Infectivity of *Cryptosporidium parvum* in Healthy Volunteers. *N. Engl. J. Med.* **1995**, *332* (13), 855–859.

Dworkin, M.S.; Goldman, D.P.; Wells, T.G.; Kobayashi, J.M.; Herwaldt, B.L. Cryptosporidiosis in Washington: An Outbreak A\associated with Well Water. *J. Infect. Dis.* **1996**, *174* (6), 1372–1376.

Elwin, K.; Chalmers, R.M.; Roberts, R.; Guy, E.C.; Casemore, D.P. Modification of a Rapid Method for the Identification of Gene Specific Polymorphisms in *Cryptosporidium parvum* and Its Application to Clinical and Epidemiological Investigations. *Appl. Environ. Microbiol.* **2001**, *67*(12), 5581–5584.

Enriquez, V.; Rose, J.B.; Enriquez, C.E.; Gerba, C.P. In *Protozoan Parasites and Water;* Betts, W.B.; Casemore, D.P.; Fricker, C.R.; Smith, H.V.; Watkins, J., Eds.; The Royal Society of Chemistry, Cambridge, U.K.:1995; pp. 84–86.

Fireman, M.; Reitemeier, R.F. Prevention of Calcium Carbonate Precipitation in Soil Solutions and Waters by Sodium Hexametaphosphate. *Soil Science* **1944**, *2*, 35–41.

Fontaine, M.; Guillot, E. Development of a TaqMan Qquantitative PCR Assay Specific for *Cryptosporidium parvum*. *FEMS Microbiol. Lett.* **2002**, *214* (1), 13–17.

Francy, D.S.; Simmons, O.D.; Ware, M.W.; Granger, E.J.; Sobsey, M.D.; Schaefer III, F.W. Effects of Seeding Procedures and Water Quality on Recovery of *Cryptosporidium* Oocysts from Stream Water by Using U.S. Environmental Protection Agency Method 1623. *Appl. Environ. Microbiol.* **2004**, *70* (7), 4118–4128.

Fricker, C.R. Detection of *Cryptosporidium* and *Giardia* in Water; In *Protozoan Parasites and Water;* Betts, W.B.; Casemore, D.P.; Fricker, C.R.; Smith, H.V.; Watkins, J., Eds.; The Royal Society of Chemistry, Cambridge, U.K.: 1995; pp. 91–96.

Gennaccaro, A.L.; McLaughlin, M.R.; Quintero-Betancourt, W.; Huffman, D.E.; Rose, J.B.. Infectious *Cryptosporidium parvum* Oocysts in Final Reclaimed Effluents. *Appl. Environ. Microbiol.* **2003**, *69* (8), 4983–4984.

Gibson III, C.J.; Stadterman, K.L.; States, S.; Sykora, J. Combined Sewer Overflows: A Source of *Cryptosporidium* and *Giardia*?. *Water. Sci Technol*. **1998**, *38* (12), 67–72.

Guyot, K.; Follett-Dumoulin, A.; Recourt, R.; Lelièvre, E.; Cailliez, J.C.; Dei-Cas, E. PCR-Restriction Fragment Length Polymorphism Analysis of a Diagnostic 452-Base Pair DNA Fragment Discriminates between *Cryptosporidium parvum* and *C. meleagridis* and between *C. parvum* Isolates of Human and Animal Origin. *Appl. Environ. Microbiol*. **2002**, *68* (4) 2071–2076.

Hayes, E.B.; Matte, T.D.; O'Brien, T.R.; McKinley, T.W.; Lodsdon, G.S.; Rose, J.B.; Ungar, B.L.P.; Word, D.M.; Pinsky, P.F.; Cummings, M.L.; Wilson, M.A; Long, E.G.; Hurwitz, E.S.; Juranek, D.D. Large Community Outbreak of Cryptosporidiosis Due to Contamination of a Filtered Water Supply. *N. Engl. J. Med.* **1989**, *320* (21), 1372–1376.

Iacovski, R.B.; Barardi, C.R.M.; Somões, C.M.O. Detection and Enumeration of *Cryptosporidium* sp. in Sewage Sludge Samples from the City of Florianópolis (Brazil) by Using Immunomagnetic Separation Combined with Immunofluorescence Assay. *Waste Manage. Res.* **2004**, *22* (3), 171–176.

Joce, R.E.; Bruce, J.; Kiely, D.; Noah, N.D.; Dempster, W.B.; Stalker, R.; Gumsley, P.; Chapman, P.A.; Norman, P.; Watkins, J.; Smith, H.V.; Price, T.J.; Watts, D. An Outbreak of Cryptosporidiosis Associated with a Swimming Pool. *Epidemiol. Infect.* **1991**,*107* (3), 497–508.

Johnson, D.; Pieniazek, N.; Rose, J. DNA Probe Hybridization and PCR Detection of *Cryptosporidium* Compares to Immunofluorescence Assay. *Water Sci. Technol.* **1993**, *27* (3–4), 77–84.

Johnson, D.W.; Pieniazek, N.J.; Griffin, D.W.; Misene, L.; Rose, J.B. Development of a PCR Protocol for Sensitive Detection of *Cryptosporidium* Oocysts in Water Samples. *Appl. Environ. Microbiol.* **1995**, *61*(11),3849–3855.

Jolis, D.; Pitt, P.; Hirano, R. Risk Assessment for *Cryptosporidium parvum* in Reclaimed Water. *Water Res.* **1999,** *33* (13), 3051–3055.

Kasper, D.R. Pre- and Post-Treatment Processes for Membrane Water Treatment Systems. In *Proceedings of the AWWA Membrane Technologly Conference;* Baltimore, MD, August 1-4, 1993; American Water Works Assoc.: Denver Colo., 1993; p. 105.

Kfir, R.; Hillner, C. Du Preez, M.; Bateman, B.Studies on the Prevalence of *Giardia* Cysts and *Cryptosporidium* Oocysts in South African Water. *Water Sci. Technol.* **1994**, *31* (5-6), 435–438.

Kramer, M.H.; Herwaldt., B.L.; Craun, G.F.; Calderon, R.L.; Juranek, D.D. Waterborne Disease: 1993 and 1994. *J. Am. Water Works Assoc.* **1996**, *88* (3), 66–80.

Kuhn, R.C.; Rock, C.M.; Oshima, K.H. Effects of pH and Magnetic Material on Immunomagnetic Separation of *Cryptosporidium* Oocysts from Concentrated Water Samples. *Appl. Environ. Microbiol.* **2002**, *68* (4), 2066–2070.

Laxer, M.A.; Timblin, B.K.; Patel, R.J. DNA Sequences for the Specific Detection of *Cryptosporidium parvum* by Polymerase Chain Reaction. *Am. J .Trop .Med. Hygiene* **1991**, *45* (6), 688–694.

LeChevallier, M.W.; Di Giovanni, G.D.; Clancy, J.L.; Bukhari, Z.; Bukhari, S.; Rosen, J.S.; Sobrinho, J.; Frey, M.M. Comparison of Method 1623 and Cell Culture-PCR for Detection of *Cryptosporidium* spp. in Source Waters. *Appl. Environ. Microbiol.* **2003**, *69* (2),971–979.

Leland, D., J. McAnulty, W. Keene; G. Stevens. A Cryptosporidiosis Outbreak in a Filtered-Water Supply. *J. Am. Water Works Assoc.* **1993**, *85* (6), 34–42.

Linden, K.G.; Oliver, J.D.; Sobsey, M.D.; Shin, G.A. *Fate and Persistence of Pathogens Subjected to Ultraviolet Light and Chlorine Disinfection;* Project 98-HHE-2. Water Environment Research Foundation and International Water Association: Alexandria, VA, 2004.

MacKenzie, W.R.; Hoxie, N.J.; Proctor, M.E.; Gradus, M.S.; Blair, K.A.; Peterson, D.E.; Kazmierczak, J.J.; Addiss, D.G..; Fox, K.R.; Rose, J.B.; Davis, J.P. A Massive Outbreak in Milwaukee of *Cryptosporidium* Infection Transmitted through the Public Water Supply. *N. Engl. J. Med.* **1994**, *331*(3), 161–167.

Madore, M.S.; Rose, J.B.; Gerba, C.P.; Arrowood, M.J.; Sterling, C.R. Occurrence of *Cryptosporidium* Oocysts in Sewage Effluents and Selected Surface Waters. *J. Parasitol.* **1987**. *73* (4), 702–705.

Massanet-Nicolau, J. New Method Using Sedimentation and Immunomagnetic Separation for Isolation and Enumeration of *Cryptosporidium parvum* Oocysts and *Giardia lamblia* Cysts. *Appl. Environ. Microbiol.* **2003**, *69* (11), 6758–6761.

Mayer, C.L.; Palmer, C.J. Evaluation of PCR, Nested PCR; Fluorescent Antibodies for Detection of *Giardia* and *Cryptosporidium* Species in Wastewater. *Appl. Environ. Microbiol.* **1996**, *62* (6), 2081–2085.

McCuin, R.M.; Bukhari, Z. Sobrinho, J.; Clancy, J.L. Recovery of *Cryptosporidium* Oocysts and *Giardia* Cysts from Source Water Concentrates Using Immunomagnetic Separation. *J. Microbiol. Methods* **2001a,** *45*(2) 69–76.

McCuin, R.M.; Hargy, T.M. Amburgey, J.E.; Clancy, J.L. Improving Methods for Isolation of *Cryptosporidium* Oocysts and *Giardia* Cysts from Source and Finished Waters. *Proceedings of the AWWA Water Quality Technology Conference*; American Water Works Association: Denver, Colo., 2001b [CD-ROM].

McCuin, R.M.; Clancy, J.L. Modifications to United States Environmental Protection Agency Methods 1622 and 1623 for Detection of *Cryptosporidium* Oocysts and *Giardia* Cysts in Water. *Appl. Environ. Microbiol.* **2003**, *69* (1), 267–274.

Medema, G.J.; Schets, F.M.; Teunis, P.F.M.; Havelaar, A.H. Sedimentation of Free and Attached *Cryptosporidium* Oocysts and *Giardia* Cysts in Water. *Appl. Environ. Microbiol.* **1998**, *64*(11), 4460–4466.

Medema, G.J.; Schijven, J.F. Modelling the Sewage Discharge and Dispersion of *Cryptosporidium* and *Giardia* in Surface Water. *Water Res.* **2001**, *35*(18), 4307–4316.

MMWR. Foodborne Outbreak of Diarrheal Illness Associated with *Cryptosporidium parvum*-Minnesota, 1995. *MMWR*. **1996**, *45*(36), 783–784.

Moon, H.W.; Bemrick, W.J. Faecal Transmission of Calf Cryptosporidia between Calves and Pigs. *Veterinary Pathol.* **1981**, *18*(2), 248–255.

Nieminski, E.C.; Schaefer III, F.W.; Ongerth, J.E. Comparison of Two Methods for Detection of *Giardia* Cysts and *Cryptosporidium* Oocysts in Water. *Appl. Environ. Microbiol.* **1995**, *61* (5), 1714–1719.

Nichols, R.A.B.; Campbell, B.M.; Smith, H.V. Identification of *Cryptosporidium* spp. Oocysts in United Kingdom Noncarbonated Natural Mineral Waters and Drinking Waters by Using a Modified Nested PCR-Restriction Fragment Length Polymorphism Assay. *Appl. Environ. Microbiol.* **2003**, *69* (7), 4183–4189.

Parker, J.F.W. *Cryptosporidium* sp. Oocysts in the Aquatic Environment: Occurrence, Removal and Destruction. Ph.D. Thesis. University of Glasgow, Glasgow, 1993.

Parkhurst, D.F.; Stein, D.A. Determining Average Concentration of *Cryptosporidium* and Other Pathogens in Water. *Environ. Sci. Technol.* **1998**, *32* (21), 3424–3429.

Payment, P.; Plante, R.; Cejka, P. Removal of Indicator Bacteria, Human Enteric Viruses, *Giardia* Cysts and *Cryptosporidium* Oocysts at a Large Wastewater Primary Treatment Facility. *Can. J. Microbiol.* **2001**, *47* (3), 188–193.

Quintero-Betancourt, W.; Gennaccaro, A.L.; Scott, T.M.; Rose, J.B. Assessment of Methods for Detection of Infectious *Cryptosporidium* Oocysts and *Giardia* Cysts in Reclaimed Effluents. *Appl. Environ. Microbiol.* **2003**, *69* (9), 5380–5388.

Quintez-Diaz, M.M.; Karpiscak, E.D.; Gerba, C.P. Removal of Pathogenic and Indicator Microorganisms by a Constructed Wetland Receiving Untreated Domestic Wastewater. *J. Environ. Sci. Health.* **2001**, *A36* (7), 1311–1320.

Reed, C.; G.D. Sturbaum; P.J. Hoover. *Cryptosporidium parvum* Mixed Genotypes Detected by PCR-Restricted Fragment Length Polymorphism Analysis. *Appl. Environ. Microbiol.* **2002**, *68* (1), 427–429.

Rimhanen-Finne, R.; Ronkainen, P.; Hanninen, M.-L. Simultaneous Detection of *Cryptosporidium parvum* and *Giardia* in Sewage Sludge by IC-PCR. *J. Appl. Microbiol.* **2001**, *91* (6),1030-1035.

Roach, P.D.; Olson, M.E.; Whitley, G.; Wallis, P.M. Waterborne *Giardia* Cysts and *Cryptosporidium* Oocysts in the Yukon, Canada. *Appl. Environ. Microbiol.* 1993, 59(1), 63–73.

Robertson, L.J.; Paton, C.A.; Campbell, A.T.; Smith, P.G.; Jackson, M.H.; Gilmour, R.A.; Black, S.E.; Stevenson, D.A.; Smith, H.V. *Giardia* cysts and *Cryptosporidium* Oocysts at Sewage Treatment Works in Scotland, UK. *Water Res.* 2000, 34 (8), 2310–2322.

Rochelle, P.; De Leon, R.; Stewart, M.H.; Wolfe, R. Comparison of Primers and Optimization of PCR conditions for detection of *Cryptosporidium parvum* and *Giardia lamblia* in water. *Appl. Environ. Microbiol.* 1997a, 63 (1), 106–114.

Rochelle, P.; Ferguson, D. Handojo, T. De Leon, R. Stewart, M.H.; Wolfe, R. An Assay Combining Cell Culture with Reverse Transcriptase PCR to Detect and Determine the Infectivity of Waterborne *Cryptosporidium parvum*. *Appl. Environ. Microbiol.* **1997b**, *63*(5),2029–2037.

Rochelle, P.A.; De Leon, R.; Johnson, A.M.;Stewart, M.H.; Wolfe, R.L. Evaluation of Immunomagnetic Separation for Recovery of Infectious *Cryptosporidium parvum* Oocysts from Environmental Samples. *Appl. Environ. Microbiol.* **1999,** *65*(2), 841–845.

Rochelle, P.A.; D.M. Ferguson; A.M. Johnson; R. De Leon. Quantification of *Cryptosporidium parvum* Infection in Cell Culture Using a Colorimetric *in Situ* Hybridization Assay. *J. Eukaryot. Microbiol.* **2001**, *48* (5), 565–574.

Rochelle, P.A.; Marshall, M.M.; Mead, J.R.; Johnson, A.M.; Korich, D.G.; Rosen, J.S.; DeLeon, R. Comparison of in Vitro Cell Culture and a Mouse Assay for Measuring Infectivity of *Cryptosporidium parvum. Appl. Environ. Microbiol.* **2002**, *68* (8), 3809–3817.

Roefer, P.A.; Monscvitz, J.T.; Rexing, D.J. The Las Vegas Cryptosporidiosis Outbreak. *J. Am. Water Works Assoc.* **1996**, *88* (9), 95–106.

Schock, M.R. Internal Corrosion and Deposition Control; In *Water Quality and Treatment, A Handbook of Community Water Supplies;* Pontius, F., Eds.; Am. Water Works Assoc.: Denver Colo.,1990; Chap. 17.

Shepherd, K.M.; Wyn-Jones, A.P. An Evaluation of Methods for the Simultaneous Detection of *Cryptosporidium* Oocysts and *Giardia* Cysts from Water. *Appl. Environ. Microbiol* **1996**, *62*(4), 1317–1322.

Skerrett, H.E.; Holland, C.V. The Occurrence of *Cryptosporidium* in Environmental Waters in the Greater Dublin Area. *Water Res.* **2000,** *34* (15), 3755–3760.

Slifko**,** T.R.; Friedman, D.; Rose, J.B.; Jakubowski, W. An *in Vitro* Method for Detecting Infectious Cryptosporidium Oocysts with Cell Culture. *Appl. Envir. Microbiol.* **1997**, *63* (9), 3669–3675.

Slifko, T.R.; D.E. Huffman; J.B. Rose. A Most Probable Number Assay for Enumeration of Infectious *Cryptosporidium parvum* Oocysts. *Appl. Environ. Microbiol.* **1999**, *65* (9), 3936–3941.

Solo-Gabriele, H.; Neueister, S. U.S. Outbreaks of Cryptosporidiosis. *J. Am. Water Works Assoc.* **1996**, *88* (9), 76–86.

Stadterman K.L.; Sninsky, A.M.; Sykora, J.L. Removal and Inactivation of *Cryptosporidium* Oocysts by Activated Sludge Treatment and Anaerobic Digestion. *Water Sci. Technol.* **1995**, *31* (5–6), 97–104.

Stinear, T.; Matusan, A.; Hines, K.; Sandrey, M. Detection of a Single Viable *Cryptosporidium parvum* Oocyst in Environmental Water Concentrated by Reverse Transcription-PCR. *Appl. Environ. Microbiol.* **1996**, *62*(2), 3385–3390. [Erratum appears in *Appl Environ Microbiol.* **1997**, *63*(2), 815].

Straub, T.M.; Pepper, I.L.; Gerba, C.P. Hazards from Pathogenic Microorganisms in Land Disposed Sewage Sludge. *Rev. Environ. Contam. Toxicol.* **1993**, *132*, 55–91.

Sturbaum, G.D.; Klonicki, P.T.; Marshall, M.M.; Jost, B.H.; Clay, B.L.; Sterling C.R. Immunomagnetic Separation (IMS)-Fluorescent Antibody Detection and IMS-PCR Detection of

Seeded *Cryptosporidium parvum* Oocysts in Natural Waters and Their Limitations. *Appl. Envrion. Microbiol.* **2002,** *68* (6), 2991–2996.

U.S. Environmental Protection Agency. *Environmental Regulations and Technology. Control of Pathogens and Vector Aattraction in Sewage Sludge*; EPA/625/R-92/013; Office of Research and Development, U.S. EPA: Washington, D.C., 1992.

U.S. Environmental Protection Agency. *Information Collection Rule. ICR Microbial Laboratory Manual;* EPA/600/R-95/178; Office of Research and Development, U.S. EPA: Washington, D.C., 1996.

U.S. Environmental Protection Agency. *Method 1622*: Cryptosporidium *in Water by Filtration/IMS/FA;* EPA/821/D-97-001; Office of Water, U.S. EPA: Washington, D.C., 1997.

U.S. Environmental Protection Agency. *Method 1622*: Cryptosporidium *in Water by Filtration/IMS/FA;*. EPA 821-R-98-010; Office of Water, U.S. EPA: Washington, D.C., 1998.

U.S. Environmental Protection Agency. *Method 1623*: Cryptosporidium *and* Giardia *in Water by Filtration/IMS/FA*; EPA 821-R-99-006; Office of Water, U.S. EPA: Washington, D.C., 1999.

U.S. Environmental Protection Agency. *Method 1622*: Cryptosporidium *and* Giardia *in Water by Filtration/IMS/FA;* EPA-821-R-01-026; Office of Water, U.S. EPA: Washington, D.C., 2001

Upton, S.J.; Tilley, M.; Nesterenko, M.V.; Brillhart, D.B. A Simple and Reliable Method of Producing in Vitro Infections of *Cryptosporidium parvum* (Apicomplexa). *FEMS Microbiol. Lett.* **1994a**, 118 (1–2), 45–50.

Upton, S.J.; Tilley, M.; Brillhart, D.B. Comparative Development of *Cryptosporidium parvum* (Apicomplexa) in 11 Continuous Host Cell Lines. *FEMS Microbiol Lett.* **1994b**, *118* (3), 233-236.

Upton, S.J.; Tilley, M.; Brillhart, D. Effects of Select Medium Supplements on *in Vitro* Development of *Cryptosporidium parvum* in HCT-8 Cells. *J. Clin. Microbiol.* **1995,** *33*(2), 371–375.

Villacorta, I.M.M.; Ares-Mazas, M.E.; Duran-Oreiro, D.; Lorenzo-Lorenzo, M.J. Efficacy of Activated Sludge in Removing *Cryptosporidium parvum* Oocysts from Sewage. *Appl. Environ. Microbiol* **1992**, *58* (11), 3514–3516.

Wallis, P.M.; Erlandsen, S.L.; Isaac-Renton, J.; Olson, M.E.; Robertson, W.J.; Van Keulen, H. Prevalence of *Giardia* Cysts and *Cryptosporidium* Oocysts and Characterization of *Giardia* spp. Isolated from Drinking Water in Canada. *Appl. Environ. Microbiol.* **1996,** *62* (8), 2789–2796.

Warnecke, M.; Weir, C.; Vesey, G. Evaluation of an Internal Positive Control for *Cryptosporidium* and *Giardia* Testing in Water Samples. *Lett. Appl. Microbiol.* **2003,** *37*, 244–248.

Water Supply (Water Qualtiy) Regulations, no. 3184 United Kingdom Drinking Water Inspectorate: London, U.K., 2000.

Watt, P.M.; Johnson, D.C.; Gerba, C.P. Improved Method for Concentration of *Giardia, Cryptosporidium* and Poliovirus from Water. *J. Environ. Sci. Health*, **2002.** A37 (3),321–330.

Wu, Z.; Nagano, I.; Boonmars, T.; Nakada, T.; Takahashi, Y. Intraspecies Polymorphism of *Cryptosporidium parvum* Revealed by PCR-Restriction Fragment Length Polymorphism (RFLP) and RFLP-Single Strand Conformational Polymorphism Analyses. *Appl. Environ. Microbiol.* **2003**, *69* (8), 4720–4726.

Xiao, L.; Singh, A.; Limor, J.; Graczyk, T.; Gradus, S.; Lal, A. Molecular Characterization of *Cryptosporidium* Oocysts in Samples of Raw Surface Water and Wastewater. *Appl. Environ. Microbiol.* **2001**, *67*(3),1097–1101.

Yakub, G.P.; Stadterman-Knauer, K.L. Evaluation of Immunomagnetic Separation for Recovery of *Cryptosporidium parvum* and *Giardia duodenalis* from High-Iron Matrices. *Appl. Environ. Microbiol.* **2000,** *66*(8)**,** 3628–3631.

Zuckerman, U.; Armon, R.; Tzipori, S.; Gold, D. Evaluation of a Portable Differential Continuous Flow Centrifuge for Concentration of *Cryptosporidium* Oocysts and *Giardia* Cysts from water. *J. Appl. Microbiol.* **1999**, *86*(6), 955–961.

WASTEWATER UTILITY

Alabama
Montgomery Water Works & Sanitary Sewer Board

Alaska
Anchorage Water & Wastewater Utility

Arizona
Gila Resources
Glendale, City of, Utilities Department
Mesa, City of
Peoria, City of
Pima County Wastewater Management

Arkansas
Little Rock Wastewater Utility

California
Calaveras County Water District
Central Contra Costa Sanitary District
Crestline Sanitation District
Delta Diablo Sanitation District
Dublin San Ramon Services District
East Bay Dischargers Authority
East Bay Municipal Utility District
Eastern Municipal Water District
El Dorado Irrigation District
Fairfield-Suisun Sewer District
Fort Lauderdale, City of
Irvine Ranch Water District
Las Virgenes Municipal Water District
Livermore, City of
Lodi, City of
Los Angeles, City of
Los Angeles County, Sanitation Districts of
Napa Sanitation District
Orange County Sanitation District
Palo Alto, City of
Riverside, City of
Sacramento Regional County Sanitation District
San Diego Metropolitan Wastewater Department, City of
San Francisco, City & County of
San Jose, City of
Santa Barbara, City of
Santa Cruz, City of
Santa Rosa, City of
South Bayside System Authority
South Coast Water District
South Orange County Wastewater Authority
Sunnyvale, City of
Union Sanitary District
West Valley Sanitation District

Colorado
Boulder, City of
Colorado Springs Utilities
Greeley, City of
Littleton/Englewood Water Pollution Control Plant
Metro Wastewater Reclamation District, Denver

Connecticut
The Mattabassett District
New Haven, City of, WPCA

District of Columbia
District of Columbia Water & Sewer Authority

Florida
Broward, County of
Fort Lauderdale, City of
Gainesville Regional Utilities
JEA
Miami-Dade Water & Sewer Authority
Orange County Utilities Department
Reedy Creek Improvement District
Seminole County Environmental Services
St. Petersburg, City of
Stuart Public Utilities
Tallahassee, City of
Tampa, City of
Toho Water Authority
West Palm Beach, City of

Georgia
Atlanta Department of Watershed Management
Augusta, City of
Clayton County Water Authority
Cobb County Water System
Columbus Water Works
Fulton County
Gwinnett County Department of Public Utilities
Savannah, City of

Hawaii
Honolulu, City & County of

Idaho
Boise, City of

Illinois
American Bottoms Wastewater Treatment Plant
Greater Peoria Sanitary District
Kankakee River Metropolitan Agency
Metropolitan Water Reclamation District of Greater Chicago
Wheaton Sanitary District

Iowa
Ames, City of
Des Moines Metro Wastewater Reclamation Authority
Iowa City

Kansas
Johnson County Unified Wastewater Districts
Unified Government of Wyandotte County/Kansas City, City of

Kentucky
Louisville & Jefferson County Metropolitan Sewer District

Louisiana
Sewerage & Water Board of New Orleans

Maine
Bangor, City of
Portland Water District

Maryland
Anne Arundel County Bureau of Utility Operations
Howard County Department of Public Works
Washington Suburban Sanitary Commission

Massachusetts
Boston Water & Sewer Commission
Upper Blackstone Water Pollution Abatement District

Michigan
Ann Arbor, City of
Detroit, City of
Holland Board of Public Works
Lansing, City of
Saginaw, City of
Wayne County Department of Environment
Wyoming, City of

Minnesota
Rochester, City of
Western Lake Superior Sanitary District

Missouri
Independence, City of
Kansas City Missouri Water Services Department
Little Blue Valley Sewer District
Metropolitan St. Louis Sewer District

Nebraska
Lincoln Wastewater System

Nevada
Henderson, City of

New Jersey
Bergen County Utilities Authority
Passaic Valley Sewerage Commissioners

New York
New York City Department of Environmental Protection

North Carolina
Charlotte/Mecklenburg Utilities
Durham, City of
Metropolitan Sewerage District of Buncombe County
Orange Water & Sewer Authority

Ohio
Akron, City of
Butler County Department of Environmental Services
Columbus, City of
Metropolitan Sewer District of Greater Cincinnati
Northeast Ohio Regional Sewer District
Summit, County of

Oklahoma
Trinity River Authority
Tulsa, City of

Oregon
Clean Water Services
Eugene/Springfield Water Pollution Control
Water Environment Services

Pennsylvania
Philadelphia, City of
University Area Joint Authority

South Carolina
Charleston Commissioners of Public Works
Mount Pleasant Waterworks & Sewer Commission
Spartanburg Sanitary Sewer District

Tennessee
Cleveland, City of
Knoxville Utilities Board
Murfreesboro Water & Sewer Department
Nashville Metro Water Services

Texas
Austin, City of
Dallas Water Utilities
Denton, City of
El Paso Water Utilities
Fort Worth, City of
Gulf Coast Waste Disposal Authority
Houston, City of
San Antonio Water System
Trinity River Authority

Utah
Salt Lake City Corporation

Virginia
Alexandria Sanitation Authority
Arlington, County of
Fairfax County
Hampton Roads Sanitation District
Henrico, County of

Hopewell Regional Wastewater Treatment Facility
Loudoun County Sanitation Authority
Lynchburg Regional WWTP
Prince William County Service Authority
Richmond, City of
Rivanna Water & Sewer Authority

Washington
Bellevue Utilities Department
Edmonds, City of
Everett, City of
King County Department of Natural Resources
Seattle Public Utilities
Sunnyside, Port of
Yakima, City of

Wisconsin
Green Bay Metro Sewerage District
Madison Metropolitan Sewerage District
Milwaukee Metropolitan Sewerage District
Racine, City of
Sheboygan Regional Wastewater Treatment
Wausau Water Works

Australia
South Australian Water Corporation
Sydney Water Corporation
Water Corporation of Western Australia

Canada
Greater Vancouver Regional District
Regina, City of, Saskatchewan
Toronto, City of, Ontario
Winnipeg, City of, Manitoba

Mexico
Servicios de Agua y Drenaje de Monterrey, I.P.D.

New Zealand
Watercare Services Limited

United Kingdom
Yorkshire Water Services Limited

STORMWATER UTILITY

California
Los Angeles, City of, Department of Public Works
Monterey, City of
San Francisco, City & County of
Santa Rosa, City of

Colorado
Boulder, City of

Georgia
Griffin, City of

Iowa
Cedar Rapids Wastewater Facility
Des Moines Metro Wastewater Reclamation Authority

Kansas
Overland Park, City of

Maine
Portland Water District

North Carolina
Charlotte, City of, Stormwater Services

Pennsylvania
Philadelphia, City of

Tennessee
Chattanooga Stormwater Management

Washington
Seattle Public Utilities

STATE

Arkansas Department of Environmental Quality
Fresno Department of Public Utilities
Kansas Department of Health & Environment
Urban Drainage & Flood Control District, CO

CORPORATE

ADS Environmental Services
The ADVENT Group Inc.
Alan Plummer & Associates
Alden Research Laboratory Inc.
Aqua-Aerobic Systems Inc.
Aquateam–Norwegian Water Technology Centre A/S
ARCADIS
Associated Engineering
BaySaver Inc.
Black & Veatch
Blasland, Bouck & Lee Inc.
Boyle Engineering Corporation
Brown & Caldwell
Burns & McDonnell
CABE Associates Inc.
The Cadmus Group
Camp Dresser & McKee Inc.
Carollo Engineers Inc.
Carpenter Environmental Associates Inc.
CDS Technologies Inc.
Chemtrac Systems Inc.
CH2M HILL
Damon S. Williams Associates, LLC
David L. Sheridan, P.C.
Dewling Associates Inc.
Earth Tech Inc.
Ecovation
EMA Inc.
The Eshelman Company Inc.
EWT Holdings Corporation
Fay, Spofford, & Thorndike Inc.
Freese & Nichols Inc.
ftn Associates Inc.
Fuss & O'Neill Inc.
Gannett Fleming Inc.
GE Betz
Greeley and Hansen LLC
Hazen & Sawyer, P.C.
HDR Engineering Inc.
HNTB Corporation
HydroQual Inc.
Infilco Degremont Inc.
Ingersoll-Rand Energy Systems
Insituform Technologies Inc.
Institute for Environmental Technology & Industry, Korea
Jacobson Helgoth Consultants Inc.
Jason Consultants LLC Inc.
Jordan, Jones, & Goulding Inc.
KCI Technologies Inc.
Kelly & Weaver, P.C.
Kennedy/Jenks Consultants
Komline Sanderson Engineering Corporation
Lawler, Matusky & Skelly Engineers, LLP
LTI Inc.
Lombardo Associates Inc.
Malcolm Pirnie Inc.
McKim & Creed
MEC Analytical Systems Inc.
Metcalf & Eddy Inc.
MWH
New England Organics
O'Brien & Gere Engineers Inc.
Odor & Corrosion Technology Consultants Inc.
Oswald Green, LLC
PA Government Services Inc.
Parametrix Inc.
Parsons
Post, Buckley, Schuh & Jernigan
R&D Engineering/Conestoga Rover & Associates
RMC Inc.
R.M. Towill Corporation
Ross & Associates Ltd.
Royce Technologies
SAIC Maritime Technical Group
Short Elliott Hendrickson Inc.
Stantec Consulting Inc.
Stearns & Wheler, LLC
Stone Environmental Inc.
Stormwater Management Inc.
Stratus Consulting Inc.
Synagro Technologies Inc.
Tetra Tech Inc.
Trojan Technologies Inc.
URS Corporation
USfilter
Veola Water North America
Wade-Trim Inc.
Westin Engineering Inc.
Weston Solutions Inc.
Woodard & Curran
WRc/D&B, LLC
WWETCO, LLC
Zenon Environmental Inc.
Zoeller Pump Company

INDUSTRY

American Electric Power
ChevronTexaco Energy Research & Technology Company
The Coca-Cola Company
Dow Chemical Company
DuPont Company
Eastman Chemical company
Eastman Kodak Company
Eli Lilly & Company
Merck & Company Inc.
ONDEO Services
Procter & Gamble Company
PSEG Services Corporation
RWE Thames Water Plc
Severn Trent Services Inc.
United Water Services LLC

Note: List as of 2/1/05

WERF Board of Directors

WERF Research Council

WERF Technical Advisory Committees

Stormwater Technical Advisory Committee

Decentralized Research Advisory Council

WERF Product Order Form

As a benefit of joining the Water Environment Research Foundation, subscribers are entitled to receive one complimentary copy of all final reports and other products.Additional copies are available at cost (usually $10).To order your complimentary copy of a report, please write "free" in the unit price column.WERF keeps track of all orders.If the charge differs from what is shown here, we will call to confirm the total before processing.

Name Title

Organization

Address

City State Zip Code Country

Phone Fax Email

Stock #	Product	Quantity	Unit Price	Total
			Postage & Handling	
			VA Residents Add 4.5% Sales Tax	
			Canadian Residents Add 7% GST	
			TOTAL	

Method of Payment:

(All orders must be prepaid.)

❑ Check or Money Order Enclosed

❑ Visa ❑ Mastercard ❑ American Express

Account No. Exp. Date

Signature

Shipping & Handling:

Amount of Order	United States	Canada & Mexico	All Others
Up to but not more than:	Add:	Add:	Add:
$20.00	$5.00*	$8.00	50% of amount
30.00	5.50	8.00	40% of amount
40.00	6.00	8.00	
50.00	6.50	14.00	
60.00	7.00	14.00	
80.00	8.00	14.00	
100.00	10.00	21.00	
150.00	12.50	28.00	
200.00	15.00	35.00	
More than $200.00	Add 20% of order	Add 20% of order	

*minimum amount for all orders

Note: Please make checks payable to the Water Environment Research Foundation.

To Order (Subscribers Only):

Log on to www.werf.org and click on the "Product Catalog."

Phone: (703) 684-2470
Fax: (703) 299-0742.

WERF
Attn: Subscriber Services
635 Slaters Lane
Alexandria, VA 22314-1177

To Order (Non-Subscribers):

Non-subscribers may be able to order WERF publications either through WEF (www.wef.org) or IWAP (www.iwapublishing.com).Visit WERF's website at www.werf.org for details.

IWA Publishing's authorised EU representative for General Product Safety Regulations is Diane D'Arras, 15 rue Duret, 75116 Paris, France, e-mail: safety@iwap.co.uk.

Printed and bound by CPI Group (UK) Ltd, Croydon, CR0 4YY
12/05/2026
02108892-0011